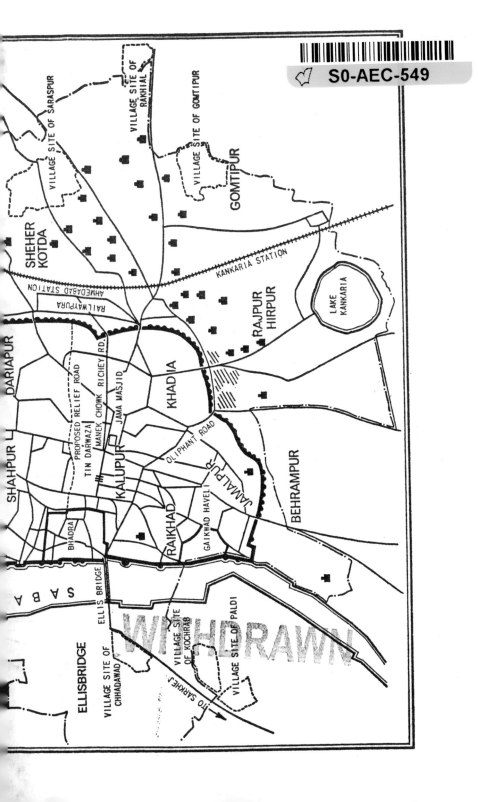

Gender Studies

MRIDULA SARABHAI

Other Books in the Series

The Veiled Woman:
Shifting Gender Equations in Rural Haryana, 1880–1990
PREM CHOWDHRY

Voices from Within:
Early Personal Narratives of Bengali Women
MALAVIKA KARLEKAR

Fertility Behaviour:
Population and Society in a Rajasthan Village
TULSI PATEL

From the Seams of History:
Essays on Indian Women
BHARATI RAY

Upholding the Common Life:
The Community of Mirabai
PARITA MUKTA

MRIDULA SARABHAI
REBEL WITH A CAUSE

APARNA BASU

DELHI
OXFORD UNIVERSITY PRESS
BOMBAY   CALCUTTA   MADRAS
1996

Oxford University Press, Walton, Street, Oxford OX2 6DP

Oxford   New York   Toronto
Delhi   Bombay   Calcutta   Madras   Karachi
Kuala Lumpur   Sangapore   Hong Kong   Tokyo
Nairobi   Dar Es Salaam
Merlbourne   Auckland

and associates in

Berlin Ibadan

© Sarabhai Foundation 1996
ISBN 0 19 563110 2

Typeset by S.J.I. Services B-17, Lajpat Nagar Part 2, New Delhi 110024
Printed by Pauls Press, New Delhi 110020
and published by Neil O'Brien, Oxford University Press,
YMCA Library Building, Jai Singh Road, New Delhi 110001

To
Nikhil and Tara

PREFACE

I had the privilege of knowing Mridula Sarabhai, her parents, her brothers, sisters, nephews and nieces. She was the eldest child of a distinguished family headed by one of India's top industrialists which also produced a renowned nuclear physicist, a poet-novelist, several social workers and dancers of repute. As far as I am concerned she was the most outstanding woman of her times that I have known. Let me elucidate.

Boss as she was known to her family and friends had really nothing very bossy about her. She was blissfully unaware of being the daughter of a wealthy mill owner; she talked on equal terms with the most powerful of politicians as she did with humble factory workers and peasants; she was more at ease sitting on broken charpoys in dhabas than on plush sofas in five star hotels. Although somewhat impatient to get on with the job in hand, she never raised her voice in anger at anyone. She was in fact a no-nonsense, serious minded woman who had no small talk or gossip. She was earnest about everything she did, come to think of it, I rarely saw her smile and not once saw her break out in guffaws of laughter. Though attractive, she took pains to make herself appear as plain-looking as she could; short cropped hair, simple, loose fitting salwar kameez of coarse khadi with a khaki dupatta thrown over her shoulders, leather chappels on her feet. Despite her small stature and nondescript appearance, Boss was a formidable person who made her presence felt wherever she was.

It was hard to believe that Mridula had been reared in the lap of luxury and had been dandled by Bapu Gandhi who looked upon her as his own daughter. As a young Congress party worker in Gujarat she had come in close contact with leaders of the Freedom movement like Sardar Patel, Acharya Kripalani and Morarji Desai. She idolized Jawaharlal Nehru who, in turn, gave her affection. Later when he became the first Prime Minister of India, Nehru turned to Mridula

and entrusted her with the dangerous task of rescuing women abducted by goondas on either side of the newly drawn borders separat-
ing Pakistan from India. She was the only woman who without any
official status could cross from one country to the other, persuade
police officials, civil servants and army personnel to accompany her
to refugee camps and homes where women were being kept in captivity and return them to their relatives. At the worst times of tension
between India and Pakistan she commanded the respect and affection
of people of both countries. She displayed a kind of fearlessness and
courage that defied description.

What made Mridula Sarabhai unique among the many women
close to the leaders of Independent India was that while others accepted rewards for what they had done by becoming ministers of the
Central Government, Chief Ministers and Governors of States, Boss
remained totally uninterested in gaining any kind of recognition for
herself. She was a living example of the principle of *Nishkama
Karma*; to her the performance of the task given to her was all the
reward she wanted.

After Independence when the communal strife in Punjab had subsided, Mridula got her involved in sorting out differences that
cropped up between the Prime Minister of India, Pandit Nehru and
the Prime Minister of Kashmir, Sheikh Abdullah. Both men were her
friends and she did her best to bring them closer to each other. It
didn't work out the way she had hoped. Sheikh Abdullah, at one time
lauded as the Lion of Kashmir for having opted for a Secular India
rather than join the Islamic Republic of Pakistan, was accused of
treason. Nehru was pressurised by his advisers to dismiss him and
put him under arrest. Mridula maintained that Nehru had been misguided by people close to him and espoused the case of Sheikh Abdullah. While Sheikh Sahib was in jail, she looked after his entire
family, organised funds for lawyers to defend him in court and tried
to rouse public opinion in India in his favour. A very reluctant Pandit
Nehru was pressurised into ordering Mridula's arrest and detention
in jail. It hurt her as much as it hurt Pandit Nehru. Strangely enough
it did not diminish their regard or affection for each other.

Mridula Sarabhai left a massive collection of correspondence she
exchanged with leaders from the time she entered public life to her
death in 1978. A sizeable portion of this including letters Mahatma
Gandhi wrote to her were in Gujerati. The Sarabhai Trust had it
catalogued and filed. I was partly responsible for persuading Gautam

and Gira Sarabhai to commission some competent author-historian to write her biography based on these documents. We were fortunate in finding Dr Aparna Basu, Gujarati wife of a Bengali Civil Servant and herself an established professor of history to undertake this arduous task. This biography is the best tribute the Sarabhai family could have paid to this truly great daughter of India.

KHUSHWANT SINGH

ACKNOWLEDGEMENT

I am deeply grateful to the Sarabhai Foundation for allowing me to consult the private papers of Mridula Sarabhai as well as to look at the correspondence of Shri Ambalal Sarabhai and Srimati Sarladevi Sarabhai with their children. I must also thank the late Shri K.M. Tripathi and other staff members of the Sarabhai Foundation, for their unfailing help and courtesy. Shri Kaliprasad, who worked with Mridulaben for over a decade, was kind enough to go through the entire manuscript and made numerous suggestions. This book could not have been written but for Gira Sarabhai. I have no words to thank her for her constant interest, warmth, generosity and hospitality. I am also grateful to all the persons (listed in Appendix I) who were kind enough to spare their time and grant me interviews. I would like to thank Dr. Nayana Goradia for persuading me to write this book, and Mr. Khuswant Singh for his encouragement and in agreeing to write the perface.

CONTENTS

LIST OF PHOTOGRAPHS

Between pages 144–145

INTRODUCTION

If I had hundred women like Mridula Sarabhai I could bring about a complete social revolution.

Mahatma Gandhi[1]

At twenty Mridula Sarabhai was already a household name in Gujarat. Stories of her unusual courage and daring during the 1930's Civil Disobedience Movement had spread throughout the province. She had courage as well as qualities of leadership. The awestruck shopkeepers used to call her *Pathan* when she picketed their liquor stores. At home her family and friends affectionately called her *'Boss'*.[2]

Jawaharlal Nehru admired her as 'an extraordinarily brave girl',[3] one of the bravest persons he had known and also as 'a tremendous worker, a great organizer with amazing grit and courage'.[4] Sarojini Naidu described her as the 'Joan of Arc' of Gujarat.[5]

Unmindful of the danger to her life, she would rush out at midnight to rescue an abducted girl or to a site where a train carrying refugees had been attacked. During the riots in 1947 in Punjab, when communal frenzy was at its height, a respectable residential area of Lahore had become like a slaughter house. Mridula wanted to rescue an abducted girl from that locality. Her colleagues asked her not to go but she said 'Let me go myself and see what it is like to be gripped by fear'. She walked alone into the area, unmindful of the violent mobs roaming around, made the necessary enquiries and returned. On the way back she picked up a one-and-a-half or two year old boy who was sitting in the middle of the road and carried him safely to a police station. She said later that the boy had been her protector and no one attacked her because of him.[6]

'She is the bravest man in the Indian army', said General Thimayya. Near the village of Shahjahanpur, in Meerut District, she had stood on a tree for seven hours pacifying an armed Jat mob from attacking Muslims. Mridula was adventurous and liked to test her ability to face danger without being daunted.

She remained fearless throughout her life. Once, when asked whether she had experienced fear, she had replied that when confronted with danger the instinctive reaction was of fear. But if one trained oneself to face danger, it was possible to function. To be non-violent and courageous called for conviction in one's beliefs.

Prominent among those who influenced Mridula's life and thought apart from her parents Saraladevi and Ambalal Sarabhai, were Mahatma Gandhi and Jawaharlal Nehru.

Mridula first met Gandhiji in 1918 when she was seven years old. Gandhiji had great affection for her—'I carried her in my arms when she was a little girl', he said in one of his last prayer meetings.[7] As the years went by, because of her family's close relationship with the Mahatma and also because of her own involvement in the freedom struggle, she grew closer to him, worked with him and imbibed many values from him, including that of fearlessness,—both physical and moral. She had the courage of her convictions to differ from Gandhiji. 'I was never his disciple, I was at a distance yet close to him.' She was to record later.[8]

Jawaharlal, she adored and hero worshipped since her teens. Jawaharlal's letters also show his deep concern for her. When Mridula was recovering from an illness in 1938, he invited her and her sister Bharati to Khali, near Almora in the Himalayas to spend a few days with him.[9] He took personal pride in Mridu, 'because to some extent, I have helped and advised her during the past few years.'[10] He was very keen that Indira should meet her and get to know her.[11] To Padmaja Naidu, he wrote that Mridula was 'quite an exceptional girl' with an amazing capacity for work.[12]

In the pre-independence days, whenever Jawaharlal went to Ahmedabad, she organized the arrangements. She visited him regularly when he was in jail in 1941 when she herself happened not to be imprisoned. In the early years of Nehru's prime ministership, she often functioned unofficially as his Personal Assistant.

Despite her deep personal loyalty and regard for Jawaharlal, she was no sycophant and did not toe the official line. She was outspoken and did not try to curry favour. If she felt that Gandhiji, Nehru or Patel were not right, she promptly and unhesitatingly told them so.

In the beginning Mridula had received a great deal of encouragement, support and personal guidance from the Gujarat Provincial Congress under the leadership of Sardar Vallabhabhai Patel and Morarji Desai. She had enjoyed a warm cordial relationship with them. She however, often found them conservative, authoritarian and undemocratic. As time went by, it was clear that her leanings were with the Congress Socialists under the leadership of Jawaharlal Nehru. Soon it was also evident that working in Gujarat would not be congenial.

Mridula was an institution builder. She had the capacity to train and enthuse workers, and inspire loyalty in them for the cause. As soon as she could hand over the running of the organization to a competent colleague she moved on to the next assignment offered to her by either Gandhiji or Jawaharlal.

Till 1946, her headquarters were at Ahmedabad and her activities were limited to Gujarat. She then moved to Delhi at the instance of Jawaharlal and was assigned a suite at the Constitution House from 1946 to 1953. Her work took her to U.P., Bihar, Noakhali, Punjab, and she frequently visited Pakistan trying to restore communal peace and harmony, rescuing and rehabilitating refugees and recovering abducted women.

Thousands of women from both sides of the frontier were rescued and rehabilitated in a span of six years. The best in Mridula emerged in these years and her capacity for work as well as qualities as a leader and organizer became established. Jawaharlal fully supported and trusted her. During this period Mridula frequently visited Kashmir and made a first hand assessment of the complex political situation that was fast deteriorating.

On the night of 9 August 1953, Sheikh Abdullah, chief minister of Kashmir and a friend of Jawaharlal as well as of Mridula, was suddenly arrested and in his place Bakshi Ghulam Mohammad was installed as chief minister. Mridula was in Amritsar when she heard this news and could not believe that Jawaharlal could have been a party to this decision.

Soon after independence in 1947, Abdul Ghaffar Khan, the Frontier Gandhi, had been put, without a trial, under indefinite detention, by the Pakistan Government and there was a great deal of criticism in India of this undemocratic act. How could India under the leadership of Jawaharlal do a similar thing in Kashmir? Mridula could not reconcile herself to such a contradiction. She suspected that Jawaharlal had been purposely misinformed about the Sheikh and the situation in Kashmir. Since assuming office as the prime minister, he was gradually losing direct contact with the masses and had started depending more and more on his official advisers and she thought that he must have been misguided. She was convinced that once Jawaharalal knew the true facts he would change his stand. But in this she did not succeed.

Mridula believed that Sheikh Abdullah was the only real, popular leader of Kashmir and it would be both politically expedient and morally correct for the Government of India to support him rather than put him in jail. She felt that a government operating from Delhi should not make and unmake governments in Kashmir: in the process it would become suspect and lose popular support in the Valley. The anti-Indian, pro-Pakistan lobby would snow ball with Sheikh's arrest. Only through an understanding with the Sheikh could India achieve a lasting integration of the State.

Though deeply interested in politics Mridula did not believe in sacrificing principles in order to get power or position.

Jawaharlal was exasperated and was not willing to accept her as the keeper of his conscience and this is perhaps where the main conflict lay between him and Mridula when it came to the question of Kashmir.

Since she could not immediately succeed at the political level to remedy the situation, in a desperate move, she tried almost single handedly, to inform and convince all those who were at all sympathetic and whose opinion mattered, in India and abroad, by putting before them the true facts.

Since the Indian Press was not willing to give her a coverage on Kashmir, undaunted, she started publishing newsletters and mailing them to people she felt would ultimately realize the wrong that was being done, and help retrace the steps which must lead to a further estrangement between India and Kashmir. Between 1953 and 1972 she published innumerable newsletters citing actual instances to make her points. She was for the integration of Kashmir with India—but her method was of winning the Kashmiris over by friendship and not by pursuing 'the stick and the carrot' tactics.

From 1953 Mridula devoted herself to Kashmir and stood by Sheikh Abdullah and his supporters even when they were accused of treason, as a result, she had to resign or leave every organization she was connected with and had to give up her room in the Constitution

House. Her phone was tapped, CID watched her house and movements, friends avoided her. There was a malicious campaign of character assassination carried out by her opponents. People told her that she was crazy as she had nothing to gain by defending Abdullah. But she did not want anything for herself and refused to give up her crusade. Although as the Congress President and the Prime Minister of India, Jawaharlal did not take the initiative of persecuting Mridula, he did not stop the Congress Party and the Home Ministry from trying to break her spirit.

She was dismissed from the Congress Party membership in 1958 and was finally detained under the Preventive Detention Act in Tihar Jail from 8 August 1958 till 6 August 1959 and later was put under House detention in her home in Ahmedabad under the Defence of India Rules. All through this period she continued to write to Jawaharlal and pleaded her point of view. From Tihar Jail, she wrote to him, 'In my personal problems, I shall go nowhere else but to you and if I am not going to be lucky or fit to get your advice, I would prefer to be self-reliant'.[13] Even while she was interned, Jawaharlal continued to accept the red rose from her, every morning, as he had ever since 1946.

The detention of Mridula Sarabhai under the Preventive Detention Act, duly attested by the Supreme Court, had aroused protests from a section of the intelligentsia who had known her since the days of the freedom struggle and had never doubted her patriotic fervour. They did not believe that she had indulged in activities which could be called anti-national and against the interest of the nation. Moreover, to keep her behind bars without a trial appeared to them undemocratic and unconstitutional. Acharya Kriplani, leader of the Praja Socialist Party in Parliament, protested against her detention without a trial.

Jawaharlal himself had never doubted her motives and her bonafides. He did not think Mridula was guilty of high treason,

but under an unfortunate set of circumstances, her courage and her capacity is being utilized and exploited for wrong and dangerous purposes. She got far greater publicity in Pakistan than in India. This is no argument, I know; but I merely say that her whole activity—not that she meant it—became so anti-national, so harmful to India that it became rather difficult to leave it where it was.[14]

The Kashmir struggle caused Mridula great emotional stress. To be treated by most of her former colleagues as a suspect guilty of treason against her own country deeply pained and disillusioned her.

First in Gandhiji's and then in Nehru's death, Mridula lost two anchors of her life. Gandhiji's assassination moved her deeply. When she heard the news in Lahore, she went to her room and wept like a child. She rushed to Delhi, and went to Rajghat.

Jawaharlal's death in 1964 was a painful blow which created 'a deep feeling of void and emotional unpheaval'.[15] Despite her opposition to Nehru's Kashmir policy, she had never lost faith in him.

Mridula had invariably got into conflict with 'the Establishment' be it the Congress Party, All India Women's Conference, Kasturba Gandhi National Memorial Trust, the United Council of Relief and Welfare. In every case although she tendered her resignation as an office bearer, rather than compromise on what was for her a matter of principle, her emotional attachment and loyalty to the institution continued. Tagore's song 'If no one listens to your call, walk on the path alone ... alone' was the one she always recalled and it gave her the strength to bear the brunt of the wrath of her friends and colleagues.

In her pre-independence photographs, we find an austere Mridula clad in simple but elegant *khadi sarees*, sometimes wearing a single flower in her long hair. She is well proportioned, sensitive and at the same time fiery. Later on Mridula can be seen in many historic photographs of post-independence days in *salwar* and *dupatta* (held in place by the gold pin, a memento from her parents), her feet clad in a pair of black Peshwari sandals, her hair bobbed closely. In photographs of her later days, one sees the effect of jail life which had ruined her health, the wear and tear of the post-Partition period, the disillusionment with old heroes turned politicians. We now notice her short stature and stocky build, an impenetrable look sometimes animated with tenderness, and a sparkle of humour. Erikson in *Gandhi's Truth* stated that Mridula had appeared in homespun *khadi*, with a dark face as 'lived in' as any you will ever have seen.[16]

All through her stormy life, her liberal and fond parents—Saraladevi and Ambalal Sarabhai—quietly stood behind her, supporting her in every way, and never interfering in the rough course she chose for herself, inspite of the anxiety it caused them. During the years she spent in prison, they never missed a single interview or letter she was allowed to receive. The warmth she received from them helped to keep up her spirits. There was a strong bond of friendship and empathy between Mridula and her parents. Although Mridula had put her work before her own life, it broke her heart to see the anxiety her parents felt on account of her.

Her wealthy father had made her financially independent almost at birth. She spent this on the causes she worked for, paying no attention to her bank balances. Her parents however made sure that her cheques were always honoured.

Mridula had from her teenage manifested certain characteristics. If she was pressurized or forced, she either resisted or withdrew. However, whenever she voluntarily accepted an assignment, she did it wholeheartedly, singlemindedly, fearlessly, honestly, following a course which she considered to be her 'inner voice'. She had empathy for the downtrodden, the exploited. She was basically 'anti-establishment' because she felt that the rights of individuals should not be bulldozed for institutional gain. Individuals had to voluntarily give up their rights for 'collective good' and bring discipline and order in their lives. She had seen the oppression of bigotry, of traditions, blindly followed. She believed in educating public opinion and not in succumbing to it under pressure. This made Mridula a marvellous friend and at the same time an opponent not to be ignored.

Upto 1953 these traits brought her fame, respect and admiration. Mridula was known for her fearless social work and her empathy for comrades in distress. She had refused offices in state and central governments in order to be free to voice her opinions. She was in complete agreement with the manifestos and official policies of the Congress and as such, as long as she was pursuing these in spirit, she considered herself a loyal Congress woman.

Throughout her life Mridula did not accept or get reconciled to the fact that cleavage between principle and practice inevitably exists in organizations. She considered it her unpleasant duty to point this out to the organization she worked for and try to narrow the difference. A defeat was never accepted as such. She would always want to try again until she succeeded. At the end more often than not, she was forced to leave the organization because the vested interest of the establishment was not amicable to fundamental change—a revolution. Her attitude caused great embarrassment to her colleagues and administrative bosses and she was accused of indiscipline and stubbornness.

Of all the prominent Congress women who took part in the freedom struggle, Sarojini Naidu, Vijayalakshmi Pandit, Rajkumari Amrit Kaur, Sucheta Kripalani, Durgabai Deshmukh—Mridula Sarabhai was the only one who did not accept an office in the National Government. She chose instead to be in the field. She had realized that a position of dependence on 'the Establishment' would be at the cost of her convictions, ideals and values. 'She is not the stuff of which ordinary mortals are made', wrote H.V.R. Iyengar[17]. She was not an arm-chair social worker but an activist to the core. 'Others talked of, philosophized about revolution, service, action, Mridula *was* it all the time'[18].

Before 1947 her main objective in life was to work towards Indian independence. After that it was for achieving communal harmony and national integration. During her last days with Gandhiji it had dawned on her that unless there was integration, freedom would be elusive.

In the last twenty years of Mridula's life, we find a conflict between colleagues, leaders and followers who had strong personal bonds of friendship and trust but who got caught up in political controversies and conflicting reports and got thrown into opposite camps seemingly inreconcilable.

Time has shown that often Mridula was right—she was however ahead of her times—she did not often succeed at the moment but was able to change the trend and success usually came decades later.

Mridula was totally dedicated to the cause she took up, 'there were no half measures for her'[19]. She placed no limit on the time she gave to the cause or the risk she was prepared to take. For twenty years she pleaded for Sheikh Abdullah. She had a stubborn capacity of remaining steadfastly loyal to the same cause irrespective of changed times and circumstances. Her one time professor, Acharya Kripalani said, 'however wrong in her opinion she may be, she was absolutely honest'[20]. 'When the faithful and the firm are so few, it does one's heart good to meet those to whom the goal is the only consideration', he wrote: 'Your firmness and determination made me look more and more within myself.... You are strengthened by a conviction nothing can shake.... I feel almost a reverence for you'.[21]

When she took up a cause, she often identified herself emotionally with those who fought for the cause, treating them as if they were her family members. In Congress, it was Jawaharlal; during the INA trial at the Red Fort it was Shah Nawaz Khan; for Pakhtunistan it was Abdul Ghaffar Khan; and for Kashmir, Sheikh Abdullah. 'This was a sad mistake', said Kripalani. Due to her emotional involvement, she was often not able to view things dispassionately.

'Some consider her a dedicated woman, dedicated to the causes which she considers noble and just; some regard her as self-opinionated and obstinate, almost perverse in her political judgement; but there is none who will deny that she has shown tremendous courage in the rough and lonely path that she has chosen for herself'.[22]

This is not an attempt at psycho-history and I have resisted the temptation of putting Mridula on a couch to speculate about the sources, motives and causes conscious and unconscious, of her drive, energy, aggressiveness or loneliness, or to go beyond what her written and spoken words warrant. This is, therefore, not so much a personal biography but the story of her public life, her work for women, for the freedom of the country, for Hindu-Muslim unity, for individual civil liberty and the right to dissent. Instead of joining the bandwagon and seeking popularity, she was usually swimming against the tide, championing unpopular causes. Mridula Sarabhai's life is worth writing about precisely because of this. From her teens she was a rebellious non-conformist. She was a product of the Gandhian age. Gandhi brought political awakening to the women of India by bringing them out of their homes and giving them a role in the Non-Cooperation and Civil Disobedience movements. His role in liberating women has been a subject of considerable controversy. A number of feminist scholars have recently argued that political participation in *satyagrahas* did not really change women's position in the family or in society as it was merely an extension of their familial role.[23] Women were assigned a specific role and retreated back to the four walls of their homes after the struggle was over. Mridula's life and activities show otherwise and throw some light on the impact of the freedom struggle on the women who participated in it, on how this gave, to many of them, not equality but a sense of self-confidence and a new image of themselves. It also reveals the attitudes of Congressmen towards the issue of gender equality.

The connection between businessmen and the Indian National Congress has been a subject of considerable historical interest.[24] The Sarabhais were a leading industrial family of Ahmedabad and this study shows the close, intimate relationship between them and the

Congress leaders. During the 1920s, 1930s and 1940s, Ahmedabad was 'the keep of Gandhism'. Industrialists like Ambalal Sarabhai and Kasturbhai Lalbhai met Gandhi and Patel frequently. They treated them with deference and followed their advice. Unlike some of their Bombay counterparts, they never opposed the Congress or even stood aloof from it when it appeared weak.

Mridula Sarabhai was involved in the major social and political movements of her time. There are many roads to the field of history, one of which is that of biography. This book is, therefore, in a way a page from the vast tome of the social and political history of modern India over the last half-a-century, seen through the public life of one person and some others who illuminated its pages.

CHAPTER ONE

EARLY INFLUENCES

Ahmedabad at the Turn of the Century

Founded in AD 1411 by Sultan Ahmed Shah, the new city of Ahmedabad was on a site close to the much older trading centre of Asawal (or Asapalli or Karnavati). It grew into a flourishing centre on the banks of the river Sabarmati, fifty miles from the river's mouth. Except in the monsoon, the Sabarmati is a thin stream in a broad bed of sand. The Sultan encouraged merchants, weavers and skilled craftsmen to come to the new capital and made it a prosperous commercial and industrial city. For a hundred years, it grew in wealth and splendour, then for sixty years it declined with the decay of the Sultanate dynasty of Gujarat and the Portuguese interference with trade. In 1572, when Ahmedabad became part of the Mughal empire and the seat of the Mughal governor of Gujarat, its prosperity recovered. In 1618, Jehangir appointed Prince Khurram (Shah Jahan) as the governor of Gujarat. The young prince ordered a Shahibag to be constructed around a new palace facing the river Sabarmati, on the outskirts of Ahmedabad. This developed into one of the best gardens of seventeenth century India.

With the disintegration of the Mughal Empire, a period of disorder set in. From 1738 to 1753 Ahmebadad was ruled jointly by the post-Mughal Muslim rulers and the Marathas. In 1757 it passed completely into Maratha hands and in 1817 with the annexation of Ahmedabad by the East India Company a new era began in its history. The old city on the east bank, covering an area of two square miles was enclosed within walls completed in 1487. The city walls which were in a dilapidated condition having collapsed in parts, were repaired by the British after 1830. At this time its population was around 80,000.

Ahmedabad, positioned between the fertile fields of Gujarat and the deserts of Rajasthan, had always been a wealthy city, its people known for their tradition of thrift and industry. It was a textile centre; brocades woven in Ahmedabad were famous throughout India and were also exported from Cambay to the Persian Gulf, ports of Arabia and South East Asia. The cooperation between Ahmedabad's Muslim weavers, Vania and Jain financiers and merchants brought great wealth to the city.[1]

When it passed over to the British in 1818, Ahmedabad had a strong business culture. Its indigenous civic and business traditions such as that of the *Nagarsheth*, the *Mahajans* and private family *Pedhi* had survived political instability of the late eighteenth and early nineteenth centuries.[2]

Society in Ahmedabad was in a state of ferment as a result of the profound economic and social changes that swept India as British commercial and political domination spread. In 1823, the Native School Book Society was formed in Bombay which printed Gujarati translations of English text books.

In 1826–7, the Government opened two primary schools in Ahmedabad. In 1844, an English school was established. By 1878, there were twentythree government schools, including one high school and two teachers' training colleges. From the 1850s, young men from Ahmedabad had started going to Bombay to study at Elphinstone College, a trend which continued to grow with the establishment of the Bombay University in 1857. Law courts encouraged the legal profession, and the law college in Bombay attracted young men from Gujarat. The East India Company's judicial and revenue services offered job opportunities to the English educated. Many of these young men who had come under the influence of new ideas from the West as well as those of the Brahmo Samaj from Bengal became interested in social and religious reforms.

Probably it was in Ahmedabad that the earliest effort for organizing Municipal self-government in Western India was made in 1833. This period also stimulated the growth of newspapers and magazines published in English as well as in Gujarati. Although *Mumbai Samachar* (Bombay News) was printed in Bombay, it had many subscribers from Ahmedabad.

The first newspaper published in Ahmedabad was *Vartman* (The Present) in 1849 by the Gujarat Vernacular Society, which had been established a year earlier, by Alexander Forbes. *Vartaman* was published every Wednesday and thus nick-named *Budhvariyu* (Wednesday). *Buddhi Prakash Mandali* (Society for Intellectual Enlightenment) started a fortnightly *Buddhi Prakash* (Intellectual Radiance) in 1850. It soon became a weekly and every issue contained articles on social reform, women's education, new discoveries in science, geography etc. The idea was to remove superstition and broaden the horizon of its readers. Three weekly newspapers were published in Ahmedabad, the Anglo-Gujarati *Hitechhu* (Well-Wisher) (1854), the Gujarati *Samsher Bahadur* (News) (1854) and *Ahmedabad Samachar Prajabandhu* (Friend of the People) (1898).[3]

The Gujarat Vernacular Society became a nucleus of the social and educational reform movements in the city and was supported by local businessmen. The British encouraged local philanthropists to start libraries, schools for boys and girls, clubs and hospitals. By 1855, Ahmedabad had a post office and the Bombay Telephone Company introduced its service in Ahmedabad in 1897 with thirty four phone subscribers.[4]

British rule thus laid the foundation of a new social and economic ethos. English education and contacts with British officials, educationists and missionaries made young Gujaratis aware of new modes of thinking and loosened the hold of traditional dogmas. A new group of public spirited citizens emerged consisting of government officials like Bholanath Sarabhai, Mahipatram Rupram, industrialists like Ranchhodlal Chhotalal and Maganbhai Karamchand, advocates like Manekchand Motichand, poets like Dalpatrai Dahyabhai and teachers like Bhogilal Pranvallabhdas and Tuljaram Mehtaji. These were men who initiated moves towards social reform and social change in the late nineteenth and early twentieth century Gujarat. Even though they belonged to different castes and religions, they collaborated for promoting new institutions. The local capitalists and intelligentsia became at once agents of change and beneficiaries of the new social and economic processes.[5]

Family Background

It was in the illustrious Sarabhai family of Ahmedabad that Mridula was born on 6th May, 1911. Her father, Ambalal Sarabhai (1890–1967) was the well-known businessman and industrialist who controlled the Calico Mills.

Ambalal's grandfather Maganbhai Karamchand (born 1823) was the head of the family firm known as Sheth Karamchand Premchand Ni Pedhi. When Maganbhai was only eleven years old, his father Karamchand had died and so Maganbhai had to take charge of the family business at a very early age. Sheth Karamchand Premchand, like several other rich merchants and bankers of Ahmedabad, financed indigenous manufactures and advanced loans to Peshwas and Gaikwads and to the chiefs of Saurashtra and Rajasthan. They jointly operated as an all-India network with offices in Bombay, Malwa, Baroda, Poona, Calcutta, Jaipur and Delhi.[6] They exported opium from Malwa to China, bought Chinese silk and tea in return and sold these to England. This triangular trade was extremely profitable and it is said that in one particular year the profits of merchants from Ahmedabad from this trade reached rupees 10 crore.[7] Leaders of the Jain and Vaishanava communities, dominated not only business but also public life.

By mid-nineteenth century, trading in opium was no longer as profitable as it had been, because of greater competition from Chinese producers and also from merchants of other cities in India. Under the East India Company's rule, the indigenous *sarafs* started losing business as bankers to army paymasters, and as money lenders to rulers of native states. The finance that became available /from this dwindling business now found its way to the mechanized textile industry with machines imported from England.

Ranchhodlal Chhotalal (1823–98) an outstanding Nagar Brahmin entrepreneur established the first textile mill in Ahmedabad in 1861. Among its large shareholders were Maganbhai Karamchand, Nagersheth Premabhai Himabhai and Hathisingh Kesarsingh. Ranchhodlal was an ex-government official of the British Raj. He was proficient in English, Sanskrit and Persian, forward looking and interested in social and religious reforms.[8]

On a visit to Bombay, Maganbhai had got acquainted with the Parsi philanthropist Sir Jamshedji Jeejeebhoy and was greatly inspired by him. Maganbhai donated Rs. 20,000 for the first girls' school in Ahmedabad which exists to this day. He also contributed towards the funding of the Gujarat College. A devout Jain, Maganbhai founded a Jain Pathshala for the study of Jainism by Jains. He donated money for building a Jain temple in Ahmedabad called 'Temple of Ashtapatiji' and another temple on the hill of Shatrunjaya

in Saurashtra. Shortly before his death, he donated a lakh of rupees
for different religious and charitable purposes.[9]

In 1864 Maganbhai died at the age of forty-one. The same year
the first railway line, the bearer of Western industrial civilization,
reached Ahmedabad from Bombay making Ahmedabad the largest
and most important commercial and industrial city of Gujarat.[10]

As Maganbhai had no son, he had adopted his daughter's son
Sarabhai. On Maganbhai's death, Sarabhai was still a minor and was
cared for by his conscientious trustees, among whom were Sir Jamshedji Jeejeebhoy, the Second Baronet and Sheth Umabhai Hathising.
The trustees also nurtured his family business.[11]

By 1876, mechanized spinning and weaving machines imported
from England had already been successfully introduced in Ahmedabad. Now the first mechanized cloth printing factory called the
Ahmedabad Calico Printing Company Limited was established as a
small venture, by eight promoters of which the firm of Sheth Karamchand Premchand was one. This business was unable to compete with
local hand printers and the Company ran into financial difficulties.
In 1880, it was taken into voluntary liquidation. The firm of Karamchand Premchand which was the largest lender, acquired the printing
works from the liquidators. A new company was floated in the same
year with a view to add cotton spinning and then weaving to the
printing works. Thus the Ahmedabad Manufacturing and Calico
Printing Company Limited came into being, which developed in
a hundred years into the thriving concern popularly known as
Calico.[12]

In 1881, Sarabhai passed the matriculation examination of the
Bombay University and joined the Gujarat College. But after about
a year, he left his studies, on his trustees' advice, to take charge of
his family business.

While Sarabhai was a Dasha Shrimali Jain of Ahmedabad, his wife
Godavariba was a Vaishnava from Surat. Godavariba had an unusual-
ly fine complexion, good looks and had a striking presence. Her
father Uttamram was a liberal. He had shifted from Surat to Karachi
and practiced law. As he was greatly drawn to Christianity his orthodox Hindu wife had decided to leave Karachi with her children
and return to Surat. That Uttamram, who was regarded as an eccentric
by his family, had actually accepted the Christian faith in his last days
was discovered on his death.

Sarabhai died in 1895 at the age of twentyeight having contacted dysentery during a trip to Kashmir. In his short life, apart from looking after the Calico Mills, he served as a member of the Ahmedabad municipality and was for some time the secretary of Maganbhai Kanyashala. In his Will, he donated a substantial sum for distribution to foundlings' homes.[13]

Only three months after Sarabhai's death, his wife Godavariba passed away leaving her nine-year-old elder daughter Anasuya, son Ambalal aged five and the youngest daughter Kanta who was still an infant. Godavariba was a great devotee of the Mata at Ambaji and had named her son Ambalal after the goddess. The three orphaned children were brought up with affection and concern by their uncle Chimanbhai, Sarabhai's younger brother, Nanubhai Jhaveri and his wife Surajba and by Sir Chinubhai Madhavlal.[14]

Sarabhai was the hereditary *Sheth* of the Dasha Shrimali Jain Community in Ahmedabad and on his death, the prestigious position passed on to Ambalal. Chimanbhai was a staunch Jain. He tried in vain to mould Ambalal to be worthy of the responsibility that he would have to shoulder, but Ambalal as well as Anasuya were nonconformist and rebellious. They were both quite clear about the different path they wanted to pursue.

Ambalal's uncle Chimanbhai died at the age of fortyfive in 1908. Ambalal became a major at the end of the same year. He immediately took over all Chimanbhai's affairs pertaining to family, caste and business in his own hands. Anasuya was a moral support on all issues.

There had been several marriage proposals for Ambalal. One of them was from Harilal Gosalia for his daughter Rewa. Gosalia was a very upright self-made man, somewhat Victorian in his outlook. An agnostic, influenced by Kant and Hegel, he was educated in Bombay and was a practicing advocate in Rajkot. His first wife had died, leaving behind three daughters. Rewa was the eldest and she helped her father in bringing up her younger sisters. Although they had to live very frugally, the fourteen year old Rewa managed the house remarkably well. She was studying in an English medium school. At their first meeting, Ambalal was very impressed by her and asked her if she would marry him of her own free will. Rewa affirmed. Rewa's ideal was Sita from the Ramayana.

Ambalal's younger sister Kanta was shy and bashful. Anasuya and Ambalal felt it their responsibility to find her a husband. Instead of looking for riches, they found a dashing young man, a mechanical engineer educated in England who had just accepted the post of the Principal of Kalabhavan (School of Art) in Baroda. That he came from a poor family was of no importance to Ambalal and Anasuya. In 1910, the marriage of Kanta took place one day before that of Ambalal. The Jain Dasha Shrimali community was offended by Ambalal's choice of partners for Kantaben as well as for himself and stones were thrown at the wedding processions.

Kanta's marriage proved disastrous. The difference in wealth became an insurmountable difficulty. Ambalal blamed himself for taking as vital a decision as marriage for another human being. From then on be did not take any initiative for any marriages—not even in the marriages of his children.

As a solution to Anasuya's unfortunate child marriage at the age of twelve, a secret trip to England for studies in medicine was planned by Ambalal and Anasuya to prevent her husband who was threatening to enforce his conjugal right. Ambalal also made Anasuya financially independent of himself.

Soon after his marriage, Ambalal moved from Shantisadan, their large joint family house in the city, to his bungalow, the Retreat, on the outskirts of the city in the Shahibag area near the Cantonment. On a nineteen acre farmland, the large bungalow had been constructed by Chimanbhai in 1904.

Ambalal would insist on taking his wife and sisters wherever he went socially. Ambalal and Saraladevi mixed easily with elders in the community, and also with British officials. The only other couple to be seen together in those days was Ramanbhai and Vidyagauri Neelkanth, both eminent social reformers. Vidyagauri was the first woman graduate in Gujarat, having pursued her studies while brining up a family.

Saraladevi continued her formal studies at Mahalaxmi Training College but had to give it up because she was expecting a baby the year after her marriage. Her passion to study, learn and to know, however, continued through her long lifetime. As she grew older and faced life, her interest in child education, Gandhian way of life, women's issues and in Hindu philosophy deepened. Both she and Ambalal made it a point to know more about these by reading, by meeting knowledgeable people in the field and endlessly discussing amongst themselves.[15]

On 6 May 1911, a daughter was born. She was delicate and so was named Mridula.

Anasuya had sailed for England early in 1911. She intended to study medicine in London, but when she realized that it would involve animal dissection, being herself a vegetarian, the very thought revolted her.[16] Among her Indian friends in London were Shri Prakasa, Manu Subedar and Dr Jivraj Mehta. They recommended her attending lectures of the Fabian Society. She came in contact with Sylvia Pankhirst and the Suffragettes who were fighting for women's right to vote. Anasuya at last felt liberated.

Meanwhile Ambalal had felt the need to learn more about the textile industry in England. He sailed for England in 1912 accompanied by Saraladevi, Mridu, his cousin Nirmala and her fiance. Bakubhai, a dashing young friend of Ambalal, was also in the houseparty. They rented a house in Richmond and stayed there for ten months.

In England, the Sarabhai family became somewhat Anglicized. Although they continued to be vegetarians, they lived in style, with an English butler, chauffeur, valet and maid. Ambalal wore suits tailored in Saville Row, tailcoat, stiff collar, top hat. Saraladevi wore chiffon *sarees*, lace blouses, corsets and adopted fashionable hairstyles. Bharati, their second daughter, was born in London. After a stay of about ten months there, they returned to India with the two English governesses who were attending to their two daughters.[1] The governesses brought up the children very strictly supervising their eating, sleeping, playing, punishing them frequently and allowing them to see their parents only at fixed times each day.

This first exposure to western culture, became a period of exploration for Ambalal, Saraladevi and Nirmala. In London, Nirmala broke off the engagement with her fiance and decided to marry Bakubhai.

Emancipated, Anasuya stayed on in England—fascinated by the stimulating life around her.

On their return from England, the Sarabhais toyed with the idea of settling down in Bombay and purchased Malden House, a three-storeyed building on Marine Lines. Furniture was bought from Wimbric and later from Benjamin, the most fashionable European furniture manufacturers in Bombay. They mixed with Bombay's westernized elite and frequented clubs and went horse-riding on *Chowpati* (a sea beach in Bombay city). Soon however, they decided to return to 'the Retreat' in Ahmedabad.

The Retreat was built on nineteen acres of farm land in Shahibag[*] in 1904 by Chimanbhai and remodelled in 1935 by Suren Kar, the

---

[*]   (Shah Jahan, when he was Viceroy of Gujarat in early seventeenth century
       ordered the laying of *Shahi-Bag* ranked as one of the finest gardens of India

designer of Rabindranath Tagore's 'Uttarayan' in Shantiniketan. The beautiful garden was adorned with replicas of Venus de Milo and other Greek and European statues around 1920; there was a rose garden, an orangery and a nursery. There were fountains and ponds with fish and lotuses. Exotic flowers, trees and bushes were planted and there were peacocks and birds of numerous kinds. In 1920, the whole family went to England for the second time. The four older children Mridula, Bharati, Suhrid and Leena accompanied the parents and there was some thought of admitting them into English public schools. The family stayed in a rented three-storeyed house in Hampstead in real style and luxury. They had two cars, an Austin and a Daimler, liveried chauffeurs, English cooks and butlers and silver cutlery monographed 'A.S.'. As Ambalal had wanted a first-hand knowledge of the textile industry in Lancashire in order to modernize Calico, he commuted frequently to Lancashire from London and finalized the purchase of fine-counts spinning machines. Calico was the first textile mill to introduce in India new technology in these fields.[18] The children were sent to a private school but the English children teased them about their brown colour and called them 'coffee'. This racialism hurt them and they refused to go to school. Miss Williams and Mr Edwin Standing, two young graduates from Cambridge and Karuna Shankar Bhatt from India were employed to educate the children at home. The family soon decided to return to India crammed with such Western ideas as seemed to their good.[19]

Schooling

It was in 1912, on the boat from England to India, that Saraladevi and Ambalal read a review in the *Times Literary Supplement* of a book by Madam Montessori on child education and upbringing which fitted in with their own ideas. There was constant conflict with the governesses about discipline and punishment and so they were sent back to England; and the parents decided to bring up their children according to the Montessori method. From now on children were included as far as possible, in all social gatherings, and accompanied their parents. Mridula started participating in all that went on around her.[20] Both Ambalal and Saraladevi were most concerned about the

---

in those days. It became in the late nineteenth century, an exclusive residential area where the mill owners *Shethias* built their palatial houses.

education of their children. They looked around in Bombay and Ahmedabad for a suitable kindergarten school for their two small children. Finding none to fit their aspirations they started looking for suitable tutors and set up the private Sarabhai school.

Among the first to join the school as a tutor was Karuna Shankar Bhatt, an educationist who was to influence not only the private Sarabhai school but also a group of intellectuals and educationists in Gujarat. Karuna Shankar 'master' taught Sanskrit, Gujarati, Bengali and aspects of Indian culture. He was liberal in his thinking and a strong exponent of *Sanskars* which he felt should be imbibed from childhood. Karuna Shankar had accompanied the family to England earlier.

Mr Standing, a graduate from Cambridge was recruited while the Sarabhais were in England and sent to Italy to get trained in the Montessori method. Standing was a teacher in the school during 1921–2. Miss Williams, also a Cambridge graduate, who had been a tutor when the Sarabhais were in England and used to take the children around, was also called and she came in 1925. Among the other tutors were the poet 'Dhumketu', the Cambridge trained mathematician, S.H. Gidwani, the writer, Valjibhai Desai, the painter, Pulin Bihari Dutt and Chhotubhai Bhatt. When the Sarabhais went to the hills for the summer, the small school, with all the staff and other paraphernalia went along with them.

Copies of letters written home by Standing as well as Williams are preserved in the Sarabhai Foundation Archives.

Miss Williams writes in 1926:

Not in books have I read, nor in dreams have I dreamed of parents so devoted to their children and to their education.... There is not a system of education in earth or sea that their parents have not read of .... In the first place, they (Ambalal and Saraladevi) decided that the children must not be deprived of a regular school routine. So within their compound, they built (1927) a fairly big house with light, airy rooms (one for each subject) and all overlooking a pleasant prospect, with flowers everywhere.[21]

Madam Montessori in her dedication to the Indian edition of *Secrets of Childhood* describes Saraladevi as the ideal Montessori mother, who dedicated herself to the cause of the child in India. Saraladevi directed the school professionally with the best available teachers, aids and equipment. The success of the experiment was obvious in the eight lively, well-informed children, eager to learn, self-confident, outgoing yet restrained. Each child was taught individually, instead of in a class. The teaching was based on the Montessori system.

The school had a full-time principal and many part or full-time tutors. A very detailed record of the progress of each pupil was maintained and tutors tried to interest the pupils in the subject rather than force them into work. The principal and teachers met Saraladevi regularly, and she and Ambalal spared no pains or money to get the best tutors available on an all-India selection basis. Sometimes, they even got tutors from abroad. There were classes of Indian music, vocal and instrumental, including *sitar, veena, dilruba,* violin, *pakhavai* and *tabla*. Two teachers from Manipur, recommended by Rabindranath Tagore, taught dancing. European music, including the piano and dancing, were taught by English teachers. The vast variety of subjects from which one could choose included Gujarati, Sanskrit, Bengali, English, History, Geography, Physics, Chemistry, Biology, Physiology and Mathematics. Besides these, there were classes in painting, sculpture, handicrafts, paper cutting, metal work, pottery, spinning, weaving and carpentry. The children performed plays, played tennis, badminton, croquet and cricket; swam, and attended riding lessons, gymnasium, archery and yoga classes;[22] they were encouraged to help in the kitchen and the garden; look after cows and the aviary; manage cash during travels and look after guests.

Ambalal and Saraladevi were broad-minded in their attitude to the teaching of religion. The children were taught the fundamentals of the various religions—Hinduism, Islam, Buddhism, Jainism and Christianity, and encouraged to celebrate religious festivals at the school.

From a very modest beginning in 1921 when Mridula, the eldest daughter was ten years old, the Retreat school, as it was called, grew and functioned till 1946 when Gira, the youngest daughter was twenty-three. Due to the difference in age between the children, there was only one student at a time in any class. The only pupils at the school were the eight Sarabhai children. Saraladevi and Ambalal felt that their approach to education was based on their own conviction and was too experimental to justify risking the future of other children. Friends often told Ambalal that it was unwise to bring up children in an ivory tower and parents were reluctant to send their children to such a 'school' for they would then become unfit to face the realities of life. Mridula, however, never regretted her schooling. She later wrote that she would would rather see children retain their sensitivity and innocence than see them become hard-boiled and 'worldly', accepting the wrongs and injustices in our society. She felt they were better prepared to face the world later on, because of their healthy and good 'foundation'. Mridula was not a particularly studious child. Karuna Shankar master was always comparing her with her younger sister Bharati who had a remarkable memory and was very quick to learn.[23] Mridula, no doubt, resented Karuna Shankar's partiality and was non-co-operative and inattentive in his classes. Even as a child, she wanted her own way and if her younger brothers and sisters did not behave according to her wishes, she would start crying. Could the assertive and dominating personality that she developed, be the result of the fact that Karunashankar did not appreciate her strong points? It was after she came under the inspiration of Gandhiji that her personality began to flower.

Gandhiji

Around 1913 Anasuya cut short her stay in England and immediately returned home on getting news that her young sister Kanta had died of meningitis.[24] The Suffragette Movement and the Labour Party programme in which she had been participating in England had motivated Anasuya to first start a crèche and then a small school for the children of millworkers with Ambalal's encouragement and support and devote herself wholeheartedly to this pioneering work.[25]

World War I broke out in 1914.

In 1915, Gandhiji on his return from South Africa, decided to settle down in Ahmedabad. It was Gandhiji's chosen home on his return to India, the place where he felt he could work best with people of his own background, speaking his own language, who were likely to be sympathetic to his ideas and visions of a new India.[26] Through Gujarati newspapers, the Ahmedabadis had learnt of Gandhiji's activities in South Africa and his *satyagraha* techniques reflecting among other things, Jain values based on non-violence and purificatory fasts. There was widespread sympathy for him and his methods. When Gandhiji was imprisoned because of his campaign against the Asiatic Registration Bill in January 1893, the Gujarat Sabha had convened a public meeting under the chairmanship of Nagarseth Chimanbhai Lalbhai, a textile millowner, to express solidarity with him. Three years later, *Prajabandhu* a local newspaper, described him as 'our patriotic countryman'. The Gujarat Sabha started taking a keen interest in the South African problem and frequently passed resolutions commending Gandhiji's work there. A massive public meeting was held in November 1913, which passed resolutions condemning the apartheid policy of the South African government. Unlike his political predecessors, Gandhiji spoke in simple Gujarati. He became a member of the Gujarat Vernacular Society, thereby identifying himself not only with the language but the organization of reformers, the men of letters and business who had founded and developed it.

In 1915 Gandhiji started the Kocharab Ashram with the support of local gentry. After this the sequence of events, that revolutionized the social, economic and political climate in Ahmedabad and Gujarat in a period of a decade, moved fast.

Gandhiji accepted an untouchable couple and their baby in the Ashram and this led to a storm of protests. Funds started drying up and various members of the Ashram threatened to leave.[27]

In his autobiography Gandhiji writes:

One morning (this must have been in late 1915 or early 1916 and thus well before the strike), shortly after Maganlal had given me warning of our monetary plight, one of the children came and said that a Sheth who was waiting in a car outside wanted to see me. I went out to him. "I want to give the Ashram some help. Will you accept it?" he asked.

"Most certainly", said I; "And I confess I am at the present moment at the end of my resources."

"I shall come tomorrow at this time," he said. "Will you be here?"

"Yes," said I, and he left.

Next day, exactly at the appointed hour, the car drew up near our quarters and the horn was blown. The children came with the news. The Sheth did not come in. I went out to see him. He placed in my hand currency notes of the value of Rs 13,000 and drove away.

I had never expected this help and what a novel way of rendering it! The gentleman had never before visited the Ashram. So far as I can remember, I had met him only once. No visit, no enquiries, simply rendering help and going away! This was a unique experience for me.... We now felt quite safe for a year![28]

This was Ambalal Sarabhai, a social rebel himself who was deeply attracted by Gandhiji. Ambalal's gift was in the long religious tradition of anonymous charity still practised by some businessmen..

This incident marked the beginning of a long and close, if ambiguous, relationship between Ahmedabad's leading mill owner and 'the saint of the spinning wheel'.

Anasuya started schools for the children of textile mill-workers the 1914 and devoted herself fully to this task. In 1916 having heard of Anasuya's schools, Gandhiji visited them. This led to Anasuya's active association with Gandhiji and the starting of a new phase in her life.[29] She increasingly came under the influence of Gandhian thought and got more and more involved with the problems facing the textile labour, adopting a Gandhian approach rather than that of the Labour Party in England. Anasuya was however too much of an individualist to want to join Gandhiji's Ashram.

In November, 1917, Gandhiji addressed a public meeting of un-touchables and Ambalal expressed his support by attending the meeting with his family. Saraladevi writes that they often visited the Ashram in the evenings and much preferred this to going to a club or socializing.

In the same year, Anasuya joined the Home Rule League and was one of the Secretaries of the Ahmedabad branch of the League.[30] In 1917—Gandhiji was forty-seven, Anasuya thirty-one, Ambalal twenty-six and Mridula a sensitive impressionable child of six.

In December 1917, the Ahmedabad textile millhands decided to go on strike over the issue of the 'plague bonus', and requested Anasuya to advise and lead them. Gandhiji was drawn into the strike because various interested parties including Anasuya ap-pealed to him to intervene. Anasuya, Gandhi and Shankerlal Banker met the workers everyday. Every morning, Gandhiji and Anasuya would address several thousands of workers from under a *babul* tree outside Shahpur Gate; Anasuya bringing him there in her car and carrying almost daily a leaflet signed by Anasuya but often written by Gandhiji, would appeal to the workers to remain peaceful, disciplined and united. Amabalal was, at the time of the strike, the President of the Mill Owners' Association and in charge of Calico Mills. He would occasionally go to Gandhiji's Ashram for a meal and Anasuya would serve him. But on the dispute about wages, the mill owners remained firm. After twenty days, the workers began to weaken because they and their families were starving, Gandhiji decided to go on a fast forcing both sides to yield. After three days, the mill owners succumbed to the pressure brought by Gandhiji and accepted the principle of arbitration and the strike ended. After the settlement, Gandhiji and Anasuya along with Ambalal were driven through the city in a 'victoria' leading a triumphant procession.[31]

The conflict between the workers and the mill owners did not create any bitterness or illwill nor did it overshadow Anasuya's and Ambalal's close relationship. Some people had suggested to Ambalal that he should stop Anasuya's allowance. Ambalal's reply was that, had she been born a son, she would have inherited half their father's property and hence he would never stop giving her what was morally, if not legally, her right.[32]

In 1918, during the Kaira *satyagraha* physical exhaustion led to a collapse in Gandhiji's health, Ambalal and Saraladevi went and brought him to their home to rest and recover.

Sheth Ambalal and his good wife came to Nadiad, conferred with my co-workers and removed me with the greatest care to his Mirzapur bungalow in Ahmedabad. It was impossible for anyone to receive more loving and selfless service than I had the privilege of having during this illness.[33]

In the same year, the Rowlatt Bill was introduced to deal with criminal conspiracies connected with revolutionary movement in India. The legislation provided for the suspension of ordinary laws, the rights and liberties of individuals. It provided for special courts, quick trials and summary punishments. The Mahatma's response to this was the formation of *satyagraha sabhas*, the members of which were pledged to disobey the Rowlatt Act which came into operation in 1919. Among the first signatories in Ahmedabad was Anasuyaben.[34]

In 1919, when a young calf in Gandhiji's Ashram suffered from smallpox and was in great pain, the Mahatma approached Ambalal for help in mercy killing. Ambalal's security men shot it. Prior to this in 1915, a rabid dog bit many others in the Calico Mill compound. The only rational way of dealing with the situation was for Ambalal to get all the dogs shot.[35] There was naturally a great uproar among the Jains. Ambalal was compared to General O'-Dwyer of Jalianwalabag. The calf incident proved to be the last straw. Ambalal resigned and was excommunicated from the Dasha Shrimali Jain caste in 1919.

E.M. Standing writes in 1921:

Mr. Sarabhai is an extraordinary mixture, being at the same time a representative of two entirely different cultures, the East and West.... Yet, at the

same time, he is a genuine Asiatic; and the mystic East is his true spiritual
home....

Once in a conversation Ambalal said:

You see, Standing, I don't like to feel myself completely immersed in business
all the day in my office. I would like to be able to get away from it all for
a few minutes to meditate on something deeper and more permanent...[36]

Mridula's imprisonment in 1930 and Suhrid's death in 1942 brought
a profound change in Ambalal. He turned to mystical enquiry and
read many books on Indian philosophy and mysticism including
the works of Sri Aurobindo, Sri Ramakrishna and Vivekanada. He
went to lectures on religion. He was an agnostic but fascinated by
the Vedanta.

In the 1930s Brij Mohan Birla had told Ambalal: 'If you want
your son to join you in business do not send him to Oxford but take
him into business when he is in his teens.' To this, Ambalal had
replied that if his children wished to pursue other interests or profes-
sions he would not discourage them, for to him, wealth was a means
not an end in itself.[37]

Ambalal was a man of versatile interests and fond of reading. He
had a vast collection of books on various subjects—philosophy,
psychology, religion, literature, biographies, history, architecture, hor-
ticulture, botany, zoology and medicine. About 50,000 books were
thus gradually collected in the Retreat library. For children,
magazines from England were often subscribed. Ambalal had an interest in birds and animals and had set up an aviary.[38]

He loved intelligent conversation and meeting talented people in
different fields. Hardly an interesting person, Indian or foreign passed
through Ahmedabad without visiting or staying with the Sarabhais.
This was especially so when Gandhiji resided in Ahmedabad upto
1931 and when several people came from all over to meet him.
Among them were the Nehrus, C.R. Das, Sarojini Naidu, Hakim
Ajmal Khan, Mohammed Ali, Shaukat Ali, Jinnah, Abdul Kalam
Azad, Dr. Annie Besant and Paul Rishard. In 1925, Rabindranath
Tagore stayed with them, not only in Ahmedabad but also for a month
in Shillong as their guest. Guests were an integral part of the ambience in which the children grew up. The Sarabhai dinner table was
the place where the family and guests met and, irrespective of their
age, participated in interminable discussions. During the 1930s and
1940s, Dr. Radhakrishnan, Sir C.V. Raman, Bhulabhai Desai,

Madame Montessori, J. Krishnamurti, Dr. Arundale and Rukmini Devi, Dilip Kumar Roy, Udaya Shankar, Prithviraj Kapoor, Harin and Kamaladevi Chattopadhyaya were among some of the guests at the Retreat.[39]

Gandhiji's personality and ideas had thus started influencing the Sarabhai family, including Mridula. Apart from her parents' close relationship with Gandhiji, Mridula's three aunts—Anasuya, Nirmala and Indumati—were also under Gandhiji's spell. By the 1930s, the *swadeshi* atmosphere had gripped the Retreat. Saraladevi presided over the Sarabhai household. Her main interest was the upbringing and education of her children but as they grew up, she took interest in public activities. She participated in the Salt *satyagraha* and advocated the use of *swadeshi* and *khadi* and was even imprisoned for a short term. She was deeply influenced by Gandhiji who also had great affection and regard for her and frequently praised in public her simplicity, humility and sense of purity and propriety.[40] He proposed her name as a trustee of the All India Spinning Association which was rejected by the other trustees on the ground that she was the wife of 'a textile king' but he asked her to 'belie the doubts and fears of sceptics' and show them that 'the wife of a mill-owner could be a lover of khadi'.[41] Gandhiji put her in charge of the Kasturba Gandhi National Memorial Trust in Gujarat.

On the night before her death, she had invited Dr. Erik Trist of the Tavistock Institute, London, for dinner and gave him an extremely clear exposition of Hindu philosophy. Her interest in children and their education, too, was with her till the end. That same night, before going to bed, she cheerfully related, to the nurse looking after her, Rabindranath Tagore's story, 'A Parrot in the Golden Cage'—an allegorical tale about how a stiff and formal education system could dull a child's joy in learning and cripple his mind and spirit. She went to bed happily, never to rise again.

There were eight children in the Sarabhai household in Ahmedabad. Mridula, the eldest, was born in 1911. Bharati was born in London two years later, then came Suhrid, Leena, Gautam, Vikram, Gita and Gira all born between 1913 and 1923. 'The Sarabhai family was like a grove of tall tress deeply linked in a family resemblance all strong and striving individuals, in their strength and their inner conflicts. True products of the individualistic upbringing they received from their father'.[42]

Thus Mridula was born, as they say, with a silver spoon in her mouth. Reared in extreme comfort and luxury, she spent her childhood secluded in the ivory tower atmosphere of the private school and the Retreat, travelling to Indian hill stations or going to Europe for her holidays. 'And then Gandhi came. He was like a powerful current of fresh air that made us stretch ourselves and take a deep breath; like a beam of light that pierced the darkness and removed the scales from our eyes; he was like a whirlwind that upset many things, but most of all the working of people's minds'.[43] Young Mridula felt the impact and rebelled.

CHAPTER TWO

SATYAGRAHA

In December 1921, the thirtysixth session of the Indian National Congress was held in Khadinagar on the banks of the Sabarmati in Ahmedabad, where the decision to launch the Non-Co-operation Movement, announced by Gandhiji in Calcutta earlier that year, was ratified. Prior to this, he had used the weapon of *satyagraha* against certain aspects of the British *Raj*, such as the land revenue structure, but in 1921, he attempted to launch Non-Co-operation against it as a total system. It was in answer to those who argued that violence alone would move the *Raj*. The chief aim of the Movement was to induce Indians to withdraw and withhold all forms of cooperation with the government and thereby paralyse it. People were asked to resign from government service, return titles, boycott law-courts, leave government schools and colleges, not wear foreign cloth or use any foreign goods. Gandhiji said that if his prescription was followed faithfully, the British would be forced to leave India within a year. This was, of course, perfectly true, but the condition set for it saved British rule for the time being at least.

Non-Co-operation was a mass movement, and it was led by a commanding personality who inspired devotion in India's millions. People experienced a new sense of freedom. Gandhiji taught them not to be afraid, to straighten their backs and raise their heads. Even in remote areas, common people talked of *swaraj* and the Mahatma.

Gandhiji moved from village to village in Gujarat, accompanied by Kasturba and Anasuyaben, appealing to women to take to spinning and weaving *khadi*. Many women, old and young, from towns and villages chose the action appropriate for them—attending meetings, staying away from schools or colleges, persuading shopkeepers not to sell foreign cloth and liquor, donating their jewellery to the Tilak *Swaraj* Fund or taking a pledge to wear *khadi*. Among those women were Anasuyaben and Indumati Chimanlal Sheth.

Mridula at this time was hardly ten years old but she attended the Congress session with her family and set up a stall to offer water to the delegates. The Gandhi *Yug* in Gujarat was passionately *swadeshi* and shaped the consciousness not only of adults but also of children who were influenced by the nationalist ambience around them. In June 1921, some beautiful silverware worth £2,300 arrived in the Sarabhai home from Mappin and Webb, London. Bharati and Suhrid were for using the dishes but Mridula, on the contrary, was quite adamantly against it. 'Although she is only ten years of age, she is already an ardent follower of Gandhi, and as a most keen nationalist will have nothing British at any price. And she is prepared, too, to make sacrifices for her ideals'.[1] She started wearing *khadi* in 1923. Mr Gidwani, a professor of Gujarat Vidyapeeth, who had taught Saraladevi Sarabhai, wrote to Mridula from Hyderabad that he had spoken to his newly married wife about her and how keen Mridula was on wearing *khadi*. In 1924, she wrote an article in *Anjali* their own children's magazine, on *khadi*. A few years later, Miss Williams also noted, 'Mridula and Bharati wear *saris*, but Mridula being a friend of Gandhi's wears homespun and sandals while Bharati wears silk and appreciates the comfort of first class English brogues for walking.'[2]

When the Prince of Wales visited India in 1921, the Congress decided to boycott the celebrations held in his honour in protest against the 'Punjab wrongs' and the Khilafat question. The Sarabhai children staged a play at home in which Mridula played the role of a young nationalist opposing the Prince's visit. Mridula's lines in the play reflected her feelings: 'The Prince is very bad. He has taken India from us Indians and has come to get a lot of money out of India.... The Prince ought to give up India now and let us have *swaraj* (Home Rule)'. Bharati on the other hand, playing the role of a pro-British Indian and taking an opposite point of view said, 'No, excuse me, it is not so. India was taken by the English long ago and it was taken from us because, at that time, we were fighting amongst ourselves and could not drive them away. So it was really our own fault'.[3]

Once, when Gandhiji was being taken to Sabarmati Ashram from the railway station in Ambalal's car, he, Miss Williams and Mridula occupied the back seat, Bharati was in the front seat. Bharati, he chaffed for clinging to her silks and satins. Mridula, he patted affectionately on the shoulder as if to say 'she and I understand one another'.[4] Miss Williams noted that Mridula was 'imbued (terribly and painfully for one so young) with the idea of doing social work and loathes government protection in anything.'[5] Being English however, she was perhaps not aware of how markedly the issue could affect the everyday life of an Indian. For instance, in 1924, Saraladevi Sarabhai was not keeping well and went to Simla with the children. Young Mridula was already wearing *khadi* and the Davico's restaurant refused to serve her tea because of this. She was extremely indignant and expressed her anger in a letter to her father.[6] Standing describing Mridula to his family in 1921, wrote that she was only ten, but she had at times an almost grown up manner.

A motherly creature, she takes great interest in the welfare of the younger members of the family, Vikram (the youngerst) specially.... She is a very serious and conscientious little person: I sometimes think it would do her good to grow younger, for a year or two, instead of growing older! She nearly always dresses Indian fashion, that is with a long skirt and a *saree* whereas Bharati prefers to dress like an English school girl in frock and tunic... Mridula wears her hair long and mostly hidden in the fold of her *saree*... Bharati has an enthusiasm for things English and is constantly plying me with questions about life in England. Mridula, on the other hand, has no love for the English Raj, or for anything English. In fact, I have a feeling that she is friendly to me not because I am English but inspite of it.[7]

In her jail diary, Mridula admitted that she was prejudiced against Miss Williams and Mr. Standing from the beginning, whereas her younger brothers and sisters, specially Bharati, Suhrid and Leena were drawn to them and attracted to Western culture and values towards which she had an aversion.

After finishing her education in the school run in their home, Mridula refused to join the government college in Ahmedabad or go abroad for further study unlike her brothers and sisters. Instead, she went to Gujarat Vidyapeeth which was established in 1921 by Gandhiji for students who left Gujarat College in response to the call for Non-Cooperation: it provided an education with a nationalist orientation.

The unique atmosphere in Gujarat Vidyapeeth in those days, as well as the special curriculum, no doubt, influenced Mridula in her formative years. Among her friends in the Vidyapeeth were Puratan Buch, Indravadan Thakore, Kapilrai Mehta and Navallal Jerajani. Puratan Buch taught her to spin on the *charkha*. From then onward Mridula used to spin regularly and there is a receipt to show that she gave two thousand yards of handspun yarn to the Charkha Sangh. At first she came to the Gujarat Vidyapeeth wearing a Calico Mill sari but soon started wearing *khadi*.[8] In the Vidyapeeth, there was a close relationship and free exchange of views between students and their professors, but discipline was also maintained.

In 1927, the students of Gujarat Vidyapeeth organized a social function in which they invited Gandhiji as the chief guest. Gandhiji sat on the dias and beside him were Ambalal Sarabhai and members of his family. Puratan Buch recalled how it was Gandhiji's day of silence so he had brought along a brief written speech. When the time came for him to speak, he signalled to Mridula and gave her his speech to read out. This incident made a deep and lasting impression on young Mridula.[9]

The Congress in its session at Lahore which was presided over by Jawaharlal Nehru in 1929, adopted the goal of *Purna Swaraj* (complete independence) and authorized Civil Disobedience, i.e., defiance of the government laws but it left the time and method of defiance wholly to the Mahatma. Gandhiji sent a letter to the Viceroy, Lord Irwin, which was a kind of ultimatum, consisting of a list of demands such as total prohibition, reduction of the rupee-pound ratio to 1 shilling 4 pence, reduction of land revenue by at least fifty per cent, abolition of salt tax, etc. The tax on salt cost the poor five *annas* per year and Gandhiji considered it the most iniquitous of levies. If the Viceroy accepted these demands, there would be no Civil Disobedience and the Congress would willingly participate in any conference. As Lord Irwin did not respond, Gandhiji announced that he would launch a Civil Disobedience movement and the weapon he chose for the non-violent struggle was salt. He announced his intention to launch a *satyagraha* by breaking the Salt Law by marching from Sabarmati Ashram to Dandi, a deserted village on the sea coast two hundred miles away.

A week before the march began, thousands attended the prayer meetings at the Sabarmati Ashram. On the evening of 11 March, when Gandhiji held his last prayer meeting, before starting on the March, there was a record crowd of ten thousand. 'I have faith in the righteousness of our cause and purity of our weapons', said Gandhiji, 'and where the means are clean, there God is undoubtedly present with His blessings. And where these three combine, there defeat is an impossibility. A *satyagrahi*, whether free or incarcerated is always victorious'.[10] The call was not just nationalistic but almost religious and moral, raising it to a mystical fervour. Ambalal's whole family met him that night at Satyagraha Ashram. According to Erikson, 'among the many well-wishers stretched out on the ground during the night' were Saraladevi, Mridula, then seventeen, Gautam, then only twelve, and Vikram, eleven.[11] That night, perhaps the only person who slept in the Ashram was the Mahatma himself. On 12 March, at 6.30 a.m. Gandhiji began his historic march to Dandi. The seventy-nine *satyagrahis* included scholars, newspaper editors, untouchables and weavers. The youngest in the group was sixteen and the oldest was the Mahatma himself, then sixty-one. Apart from Gujaratis there were men from Andhra, Bengal, Maharashtra, U.P., Tamilnadu and other provinces and they came from different religious communities—Hindus, Muslims, Christians. The people of Ahmedabad turned out in thousands to cheer him and for miles the roads were watered and strewn with green leaves, and flags and festoons gave an appearance of a festival. On reaching Dandi, the *satyagrahis* showed their defiance of the law by making salt illegally. Sea water was collected in big cauldrons and boiled. But no women were included in this first batch.

When the women at Sabarmati Ashram suggested that Gandhiji should take at least four or five of them along, he refused, saying that just as it would be cowardice for Hindus to keep cows in front of them while going to war, similarly it would be considered cowardly to keep women with them on this march. But women refused to be restrained in this manner. Khurshedben Naoroji, the grand-daughter of Dadabhai Naoroji, wrote an angry letter asking why he was preventing women from joining him. Mridula, then a student at Gujarat Vidyapeeth, jumped into the struggle despite the Principal, Kaka Kalelkar's orders not to do so.[12]

1930 and 1931 were years of tumultuous political agitation in India. The Dandi March was the most spontaneous, widespread and intense of all the Movements launched by Gandhiji. As Gandhiji marched to Dandi, women were everywhere on the way to greet him and hear him speak. At Abhrama on 10 April 1930, in an audience of 5,000 there were 2,000 women. At Matwad, the next day, a quarter of the audience were women. At Dandi, where he arrived on 13 April, 500 women received him.[13] The same day 450 volunteers from Ahmedabad marched to Dholera Creek to break the Salt Law. The *satyagrahis* were accompanied by 10,000 villagers. Women in hundreds were part of the crowd. After the *satyagrahis* had broken the Salt Law, the crowd, including women, beseeched Balwantrai Thakore to permit them also to break the Salt Law.[14] On the last day of the Salt March, Sarojini Naidu joined it at Dandi, and was the first woman to be arrested in the salt *satyagraha*. In his speeches all along the way from Ahmedabad to Dandi, Gandhiji exhorted women to picket shops selling liquor and foreign cloth, to wear *khadi* and spin on the *charkha*. On the day of his arrival at Dandi, he convened a conference of Gujarati women and invited among others, Saraladevi Sarabhai to attend this. At this meeting, it was resolved that women should picket shops selling liquor and imported cloth. 'If non-violence is the law of our being', he said, 'the future is with women'. In an article in *Young India* he wrote: 'Let the women of India take up these two activities, specialise in them, they would contribute more than man to the national freedom. They would have access to power and self-confidence to which they have hitherto been strangers'.[15] They should urge the owners of shops selling foreign cloth and liquor to stop their trade, close down their shops and also persuade consumers not to buy from these shops. He asked women to come out of their homes, and be fearless. A committee was formed with Mrs. Hamida Tayabji as president and Mituben Petit as secretary to implement this programme.

Boycotting foreign cloth and picketing shops selling liquor became the principal tasks assigned to women. To popularize the boycott of foreign cloth, a *Videshi Kapad Bahishkar Samiti* was formed in Ahmedabad whose president was Saraladevi Sarabhai and Mridula was its secretary. It organized almost daily processions in which women wearing saffron coloured *khadi sarees* with volunteers' badges pinned on, singing patriotic songs, marched through the streets of Ahmedabad. They collected foreign cloth and made bonfires of them. They also distributed cyclostyled sheets throughout the city.[16] The Rastriya Stree Sabha, a women's organization, launched an intensive campaign for the propagation of *swadeshi*. Volunteers went from door to door securing signatures for the pledge of *swadeshi*. A picketing association was formed in Ahmedabad, the main centres being Maskati Market, Panchkuwa Market and Ratan Pole, whose shops sold foreign cloth. In addition, godowns where foreign cloth was stored were also picketed[17]. Women volunteers stood outside shops selling foreign cloth and persuaded customers not to buy. Near the godowns, trucks carrying foreign cloth were stopped. In this work, prominent citizens of Ahmedabad, women from the families of Congressmen, as well as ordinary men and women took part. A committee was formed with Sharda Mehta as president and Mridula and Indumati Chimanlal as secretaries and women carried on their work despite anti-picketing ordinances.[18] They were full of excitement, optimism and enthusiasm. There was some danger as liquor shops were often surrounded by violent men but there was also an element of fun involved. The government policy was not to arrest women unless forced to do so as it would give undue importance and would tend to encourage the participation of 'ladies from good families.' The police were asked to exercise 'the greatest patience and restraint' and avoid use of force as far as possible.[19] If arrested in the morning, women were usually released in the evening or put in police vans, taken outside the city limits and freed. Their family members would go and receive them and carry food for them. Though few were given light imprisonment, there was no serious danger involved and the young girls must have enjoyed the publicity and adventure. *Vanara Senas* were formed by young girls and boys. They obtained information regarding shops which were carrying on clandestine sale of foreign cloth and informed the picketers of the arrival of trucks carrying foreign cloth or of the police. Mridula was one of the first to organize a *Vanara Sena* of children in Ahmedabad. Children and women organized *prabhat pheries* in the early mornings which moved round the city singing patriotic songs to the accompaniment of drums, bugles and *manjiras* (cymbals) and these became immensely popular. *'Dandi darya kinare Mohan mithun banave'* (On the sea shores of Dandi, Mohan makes salt). *'Danko vagyo ladvaiya shura jagajo re'* (the bell has sounded, brave warriors awake); *'jan jaye to java dejo par tek no khojo re'* (if you have to give up your life, do so, but do not give up the cause) as these lines reverberated through the cities and villages of Gujarat, they brought a new political awakening, specially among women.

In order to make the programme of boycott of foreign cloth more successful, the President of the picketing association, Saraladevi Sarabhai, sent a circular letter to all the mill-owners and traders, asking them not to use imported cotton yarn, artificial silk yarn or wool. She urged the traders only to sell *swadeshi* cloth. In those days there were no shops selling *khadi* and so women carried bundles of *khadi* cloth from house to house on Sundays and holidays. It was decided that *khadi* should be sold at a fixed place and an appeal to this effect was made to Sardar Patel, who, in turn, gave some money. Out of this grew the first Khadi Bhandar in Ahmedabad, which was started by Mridula's aunts Indumatiben and Nirmalaben.

In all these activities, Mridula took an active part from the very beginning. She was involved both in the picketing of liquor shops as well as in the boycott of foreign cloth. Once, news came that a large quantity of liquor was being brought into Ahmedabad to be taken to Shahibag police station and auctioned in retail shops. It was decided that women volunteers should prevent this. As liquor carts were likely to come early in the morning, the volunteers assembled in Sheth Mafatlal's bungalow the previous evening. At dawn, batches of women volunteers went out to prevent the liquor carts from coming but they were arrested before they could act and were taken to jail. News came that near Delhi Darwaza, Mridula, in her characteristic, fearless manner, had caught hold of the reins of the horses of a cart carrying liquor, prevented it from advancing and had been arrested.[20] Many young girls were thrilled by Mridula's courage and dare-devilry and were inspired to join the *satyagraha.*

For the first two years, the Movement for boycott and picketing was carried on with such enthusiasm in Ahmedabad, that hundreds of people stopped buying foreign cloth and shopkeepers were afraid to display imported goods. This Movement would not have succeeded but for the cooperation received from the mill-owners, the traders as well as the mill-workers. Nearly seventy new mills were about to be started with imported machinery, but as a result of nationalist protests, the mill-owners agreed not to do so. They also stopped the import of foreign yarn and artificial silk.

Sale of contraband salt collected from Dandi and Dharasna was taking place all over Gujarat. In Ahmedabad, batches of women moved about selling salt, singing in chorus, 'we have broken the salt law which will wreck the British Empire',[21] and crying 'Holy Salt', 'Gandhi Salt', 'salt that will free India, come and buy'.[22] Salt from Dandi and Dholera was distributed among volunteers, including Mridula and Khurshedben, each of whom got a packet of one *tola* which was sold often for Rs 500.[23]

On one occasion, when there was picketing outside Gujarat College and the police arrived, Mridula, without any hesitation, joined the boys in holding hands together to form a human chain. She organized a group of students and young people to hoist the National tricolour on the Ranchhodlal Chhotalal High School and the Government

Girls' High School, Ahmedabad. In the police lathicharge that followed, many were injured.

In May 1930, in Viramgam, the police surrounded a group of *satyagrahis* for several hours. In the scorching heat of the sun, they were naturally very thirsty. A procession of women, carrying water pitchers on their heads to give water to the *satyagrahis* was disbanded by the police. Neil and Flectcher, two Englishmen in charge of the Kharaghoda salt works and Gholap, the district magistrate, ordered troops on horses to disperse the women. Seven hundred women were lathi-charged and caned.[24] Gandhiji asked Mahadev Desai to investigate fully into this incident. Among the group of lawyers and citizens who went to investigate were Ambalal Sarabhai, Bhulabhai Desai, Bhogilal Lala, Saraladevi, Anasuyaben, Maniben Patel and Mridula.

In 1934, the negotiations between Gandhiji and the Viceroy having reached a deadlock, the Congress decided to resume the struggle. Mass Civil Disobedience was, however, suspended and in April that year, Gandhiji withdrew Individual Disobedience for everyone but himself. He asked all Congressmen to suspend Civil Resistance for *swaraj* and leave the task to him.[25] Mridula was deeply disturbed by this decision; she could not understand how the path of Civil Resistance was permissible only for one individual. Outspoken and courageous as she was, she was not afraid of seeking and expressing her reservations even to Gandhiji. She asked him the reasons for launching individual *satyagraha*, the means for strengthening it, when to suspend it or stop it, the qualities and attributes of a *satyagrahi* and the role of constructive work. She asked, 'The Congress launched *satyagraha* and civil disobedience. Is it not necessary to consult the Congress Working Committee before you take the decision that you alone will take upon yourself the responsibility of civil resistance?'[26]

In 1934, she was elected as the sole woman delegate to the Gujarat Provincial Congress Committee. When Gandhiji formally left the Congress in that year, Mridula asked him whether she should also resign, but he advised her not to do so. In 1941, when Gandhi once again launched his individual *satyagraha,* Mridula was confused about what was the path for her and asked so many questions in a long letter, that Mahadev Desai, in his reply, jokingly noted Gandhiji's remark that if she had to ask so many questions she should not be given the right to offer *satyagraha*.[27]

By 1942, Hitler had captured almost all the countries of Europe
and was advancing towards Moscow. On the Pacific Front, the
Japanese army had captured Hong Kong and Singapore and the
British had started withdrawing from Rangoon. The Japanese invasion of India seemed imminent. Gandhiji appealed to the British
to withdraw from India and leave her to her fate. This appeal was
popularly known as the 'Quit India' slogan, which created a stir in
the country and people anxiously waited to see what kind of action
was contemplated by the Mahatma. The Congress Working Committee met at Bombay on 8 August 1942, and passed a Resolution 'to
sanction, for the vindication of India's inalienable right to freedom
and independence, the starting of a mass struggle on non-violent lines
on the widest possible scale so that the country might utilise all the
non-violent strength it had gathered during the last twenty-two years
of peaceful struggle. Such a widespread struggle would be 'under the
inevitable leadership of Mahatma Gandhi.'[28] Anticipating that if the
Congress leadership was removed by arrest, 'every Indian who
desires freedom and strives for it must be his own guide.' 'Let every
Indian consider himself to be a free man .... Here all going would
not do Gandhiji declared in his passionate 'Do or Die' speech the
same day. Gandhiji and all the members of the Congress Working
Committee were immediately arrested on 9 August. The Mahatma
was arrested along with Kasturba, Mahadev Desai, his private
secretary, and Sarojini Naidu. All of them were taken to the Aga
Khan Palace at Poona, while the members of the Working Committee
were taken to Ahmednagar and lodged in the old fort. As news of
the arrests spread, there were spontaneous demonstrations, peaceful
and violent. In Ahmedabad, women took out processions and *prabhat
pheries,* held meetings, distributed secret underground leaflets, organized strikes and pickets outside schools, colleges and government
offices.

Many prominent Congress leaders like Jayaprakash Narayan,
Aruna Asaf Ali, Usha Mehta, Ashok Mehta, Ram Manohar Lohia,
Achyut Patwardhan went 'underground' and tried to keep the Movement alive. It took the authorities more than a week to trace Mridula.
She was finally arrested at Bombay on 20 August 1942. On being
released from jail in December 1943 after a period of one and a half
years, she established links with the underground movement and took
a keen interest in the welfare of the underground workers and leaders,
for whom Bombay had become the centre. She kept contact with

almost all the different groups, especially those from U.P. and Bihar. She met them in groups or individually from time to time. Through her family connections with wealthy industrialists and businessmen, who were nationalist in outlook, she sought financial help for the underground workers. She did her liaison work in such a quiet way that the government had no inkling of it.

The underground workers could not stay at one particular place for long. In order to avoid detection, they had to shift from place to place, often at very short notice. Mridula was one of those who constantly helped them by seeking assistance from courageous and reliable friends. A large number of underground workers relied on her in an emergency to find them some accommodation. During these years, special unit of CID personnel were deputed by the provincial governments to trace these fugitivies in their provinces. This was done at regular intervals. Mridula used to get advance information about such moves and she alerted the underground workers so that they could escape arrest. By 1944, when the financial position of many underground workers became bad, she extended monetary help to them. She also arranged for the employment of some under fictitious names, as they were in the government list of absconders.

The top floor of her father's Malabar Hill residence in Bombay, Kashmir House, was the centre of brisk underground political activity. Achyut Patwardhan, Mohan Lal Gautam, Aruna Asaf Ali, Madan Mohan Upadhyay (the younger brother of Shiv Datt Upadhyay, then the private secretary of Jawaharlal) and others used to meet there frequently. On his release from jail, when Nehru visited Bombay for the first time, Mridula arranged a meeting with the underground workers of U.P., Bihar and other places, at Kashmir House. He praised the excellent work done by her for the underground workers since her release from jail.

After escaping from the British Jail in Calcutta, Subhas Chandra Bose reached Germany. Bose had been close to the Sarabhais and was specially fond of Bharati and Mridula. He sailed from Germany to Japan in a German submarine during World War II in 1943 and organized the Indian National Army (INA) with the assistance of Japan. The majority of the people in INA. were those army men who had surrendered to the Japanese Army at the defeat of the British Army in Hongkong and Singapore. The headquarters of the I.N.A. was Singapore. It fought the British Army in Burma and on the Eastern Sector of India. They were able to capture Dibrugarh and Imphal from the British Army but could not advance further as assistance from Japan had ceased. Japan which was in any case facing a defeat had to surrender to American Forces after atom bombs were dropped on Hiroshima and Nagasaki. In turn, the INA had also to surrender to the British Army. All officers and soldiers of the INA were prosecuted by the British. At the Red Fort in Delhi, the trial of three officers, Col. Shah Nawaz Khan, Captain Sehgal and Lt. G.S. Dhillon was started which created a great deal of excitement throughout India. Mridula collected funds for the defence of the INA prisoners. An INA Relief Committee was set up, headed by Sardar Vallabhabhai Patel and Mridula was appointed the Secretary of the INA Relief Committee of the Gujarat Provincial Congress Committee. She took keen interest in collecting funds and providing employment to the released prisoners of the INA.

The atmosphere in the 1920s, 1930s and 1940s in Ahmedabad was emotionally charged and one could hardly escape the infection. Nirad Chaudhuri, no admirer of Gandhiji, describes the intensity of the Movement in Calcutta: 'I had started work in my new office, and, going home by train in the afternoon, I heard and saw what might be described as the *son et lumiere* of the Movement. In every square large crowds stood in serried ranks, listening with excited gestures and shouts, to the harangues of the leaders. There could be no mistaking the enthusiasm'.[29] Mridula greatly admired the revolutionary movement in Bengal, 'till then the Congress knew only one method, petitioning'. But then Gandhiji began his fight for *swaraj*, 'He evolved a new technique, non-violent, non-cooperation, civil disobedience.... Gandhiji was the Commander-in-Chief of the freedom struggle. So we wanted to follow him'.[30]

CHAPTER THREE

IN PRISON

Mridula spent several years in jail during the 1930s and 1940s. These were years of loneliness but also of introspection when she could reflect and comment on her own life. She could probably feel herself unfold better when away from her home and family. Mridula's first experience of jail life was in 1930 when she, her mother and Khurshedben were arrested, picketing shops selling foreign cloth. They were put in Sabarmati jail for three weeks. She kept a jail diary in Gujarati, in which she wrote,

We begin our day at 5.00 a.m. with morning prayers, singing *jhanda uncha rahe hamara* and end the day with the evening prayer singing *Vandemataram...* The toilets in Sabarmati jail were dirty and I was put in charge of keeping them clean. Khurshedben, therefore, called me G.O.C. of the W.C.! Vallabhbhai, Morarjibhai and Mahadev Desai came to visit us in jail.[1]

After three weeks the Gandhi-Irwin Pact was signed and all political prisoners were released. Gandhiji went to England to attend the Second Round Table Conference which unfortunately failed; even before he returned to India, Lord Willingdon ordered the arrest of Nehru, Khan Abdul Gaffar Khan and other Congress leaders as a preventive measure. Within a week of his landing in Bombay on 28th December 1931, Gandhiji was in jail and Civil Disobedience was resumed: the Indian National Congress was practically outlawed and the Gandhi-Irwin Pact had gone to pieces.

In Ahmedabad, the local Congress chose a number of successive 'dictators', so that if one got arrested, the next person could assume the leadership. Manilal Kothari, the first 'dictator' was arrested followed by Lilavati Desai and Vijayalakshmi Kanuga, the second and the third dictators respectively. Mridula, the fourth in line, was arrested on 8th January 1932, and taken once again to Sabarmati jail. She was sentenced to six months' imprisonment and fined Rs 300. There was no classification of political prisoners before 1930; but as a result of the fast unto death of Jatin Das in Lahore jail, in protest against the bad treatment meted out to prisoners in Punjab, government accepted the demand that political prisoners be lodged separately and formed three classes for political prisoners, 'A', 'B' and 'C'. Prisoners in 'A' class were free to use their own clothes and had the right to get food from outside. They were supplied with milk and butter from the jail and also had the benefit of other conveniences like cots and bedding. The food for 'B' class prisoners was the same as that supplied to 'A' class ones, but they could not get food from outside. The food given to 'C' class prisoners was of the lowest quality and was the same as that of criminal prisoners. Mridula, together with Lilavati Desai and Vijayalakshmi Kanuga was put in 'B' class, whereas Maniben Patel and Mithuben Petit were made 'C' class prisoners. In sympathy with the latter, all the 'A' and 'B' class prisoners in Sabarmati jail decided to forego their food and accept only 'C' class prisoners' food. This practice had been started by Sardar Patel in March 1930 when he was imprisoned. Eighty to ninety per cent of the political prisoners were in 'C' class and they resented this discrimination. Sardar, therefore, decided that all political prisoners should opt for 'C' class. When the jail authorities refused to listen to this, the prisoners went on a hunger strike. After sixty hours, the authorities gave in to their demands. Twenty days later, Mridula and sixteen of her fellow women prisoners were removed from Sabarmati jail to an unknown destination.

Mridula had never carried her own bags or bedding. For the first time, she had to do so as the police escorted them in the middle of the night to a special carriage waiting for them in the station yard. In one of the compartments, she saw a number of political prisoners, among whom she recognized Khandubhai Desai the labour leader from Ahmedabad. As the train passed the Retreat, she saw the *chowkidar* and thought to herself, he could little imagine that she was in this train and had seen him in the dark. This was the first time that she travelled in a third class carriage and slept at night on a wooden bench in the train. But all of them were in high spirits and full of excitement. They had tea and snacks at Anand and Baroda stations and played games on the train.

From Sabarmati, they were moved to Yeravada prison, Poona. Jail life was a totally new experience for them, so to begin with, as the iron gates clanged behind them, they were tense and excited and the novelty of prison life gave them a sense of exhilaration. There were

many new things Mridula had to learn—wash her own clothes, have
a bath in the open, sweep floors, clean utensils, cook, sew, etc. 'Today
is the first time that I bathed in the open, I do not know how to have
a bath with my clothes on. I have never done this and must confess
that I was totally unsuccessful'. She noted in her diary:

After having tea, I went to wash my *saree*, it was a new *saree* given by the
jail authorities. I put soap and tried to wash but I am afraid I have never done
this before. A fellow woman prisoner seeing me, came forward and washed
the *saree* for me. Today she did it, but I cannot ask her help every day. I
must learn to wash clothes myself.

Prison meant sleeping on the floor in the same cell with ten or twelve other
prisoners. If you turned sides at night, you touched another prisoner's body.
There was no privacy at day or night.[2]

Within five days there was an order, once again transferring her,
together with six of her fellow women prisoners, from Yeravada jail
to Belgaum. They were given only half an hour's notice to get ready.
It meant packing again, bidding goodbye to the other prisoners, being
taken to the railway station and boarding a train. The seven prisoners
and five policemen got into a third class compartment. For eating on
the way, they were given thick *chapatis,* garlic *chutney* and raw
onions and some rough blankets to cover themselves at night. Unfortunately, the prisoners did not meet any one at Poona station to
whom they could convey the news about their transfer. On arrival at
Belgaum station, there was no police van to take them and so they
sat on the ground near the station for half an hour before a closed
police van arrived and took them to the jail. There they had to wait
for another hour and a half till the jailor took them to the women's
ward which they entered through a narrow door: on the right side
there were two separate cells for 'B' class prisoners and on the left
a barrack for 'C' class prisoners. Whenever any new prisoner arrived,
the prisoners already inside were eager with curiosity. In comparison
with Sabarmati and Yeravada she found the Central Jail at Belgaum
dirtier, though the toilets could flush and the water looked clean.
There were bugs in the beddings, mosquitoes and insects crawled on
the floor of the cells. Being a light sleeper, these caused her many
a sleepless night. The 'B' class prisoners had two small rooms with
their own toilets and lavatories, while the 'C' class prisoners were
kept in a long barrack. In one of the rooms for 'B' class prisoners
were Kamaladevi Chattopadhyaya, Naju Wadia, Jhaverben Jamnadas,
Brijkumari and others. In the other room were Nanduben Kanuga,

Lilavati Desai, Maniben Patel, Lilavati Munshi, Manorama Joshi, Mridula and Devyani Desai with her small daughter. Mridula became the leader of her room.

Devyaniben Desai recalls how she was impressed by Mridula's personality, modesty and sympathetic behaviour towards her fellow prisoners:

When the jail superintendent came on his daily rounds, if anyone had any complaints or requests, Mridulaben would convey them to him, Mridulaben, Manorama Joshi and I became very good friends. Mridulaben made arrangements for extra milk and sweet limes *(mosambis)* for my daughter. Whenever any fellow prisoner was sick or unwell, she would wash her utensils and clothes. She never made anyone feel that she came from a wealthy family, nor was she at all proud.

We had to bathe in ice-cold water on early winter mornings in the open courtyard. I caught a chill and had severe pain in my lungs. I had no woollen clothes, so Mridulaben obtained permission from the jail superintendent, got some woolen *khadi* material and had a double-breasted jacket stitched for me.

When my daughter became two years and nine months old, she had to be taken away from jail as per rules. I was naturally very sad to part with my baby daughter but Mridulaben stood by me and consoled me.

I had very long hair and whenever I sat down to comb my hair, Mridulaben would jokingly come with a pair of scissors and ask me if I wanted my hair cut. Whenever she wanted her hair trimmed, she asked me to do it.[3]

Her days in prison began early. Usually the bugle blew at 5.30 a.m. to wake up the prisoners. The day began with prayers- 'Even though I was sleepy, got up at 5 a.m. to have a bath. Brushed my teeth. After the doors opened, rolled up my bedding, could bathe a second time and the long morning which I spent in reading and in doing some embroidery.'[4] They were locked up inside their cells by 7.30 p.m., sometimes earlier and spent the evenings talking and reading. When new prisoners came, they brought news from outside. A typical prison day was spent in reading, writing, spinning, sewing, talking to fellow prisoners. Both incoming and outgoing letters were censored by prison authorities.

Mridula read a great deal in prison. Among the books she read was Joan Conquest's *The Naked Truth* which made some shocking revelations about London slums and moved her deeply. She took extensive notes from the book and started wondering whether the conditions of Indian factory workers were equally sordid. She reflected on the inequality in Indian society, on charities—the rich trying to ease their conscience by giving their torn and discarded clothes to the poor. After reading this, she noted in her diary, 'The Ahmedabad mill-owners cannot see the growing discontent among the workers or they do not wish to see it'.[5] She also read C.F. Andrew's *Christ in the Silence* and *Son of Man* which gave her 'a lot of food for thought' and influenced her. She read Manning's *Confessions and Impressions* and reflected on the importance of sex education. She noted that she first learnt about sex from one of the servants in their home and what she heard shocked her. When she later associated this act with marriage, it aroused in her a revulsion for sex and marriage.[6] Among the other books she read were Isadora Duncan's *My Life; God's Gold* by John Flyman; *Being and Doing,* a biography of Lady Byron by J.M. Marriott; Goethe's life; *Bhagvad Gita; Treasure of Heaven;* Arnold's *Light of Asia; The Ant People* by Hans H. Ewere; *The Best One Act Plays of 1931;* Bartlett Burleigh Jones' *Woman of England;* and Radhakrishnan's *Hindu View of Life.*

She decided to ask for 'C' class food which consisted of *rotis* made of *bajra* or *jowar,* which often contained some grit in them; a vegetable made of pumpkin or radish or often rotten spinach; *dal* every alternate day and a quarter pound of jaggery or *gur* on Sundays. Early each morning a decoction of *jowar* flour, salt and water was dished out for breakfast. The same food was served for lunch and dinner. No milk or butter was served. In fact, all five of the women prisoners from Gujarat decided to opt for 'C' class food. Mridula had been a delicate child and therefore was given the most nutritious and hygienic food at home. She did not like the *chapatis* made of *bajra* nor *arhar dal* and lost weight. In fact, in her jail diary she confessed, that though she would not like to admit it, she never ever got used to Indian food. Jail food, not surprisingly, affected her digestion and liver. She suffered from constipation, dyspepsia, acidity and various other health problems. She lived on tea, bread, yogurt and a few other items. "I am not worried about my health. But I must be careful because if I fall sick, Mummy and Papa will worry... I must be well when I am released, so that I can get back to work quickly'.[7]

Those who were ill, on producing a slip from a doctor, could get some milk. Maniben Patel got some milk daily and often forced Mridula to have some. Once, Mridula was having *jowar rotis* with *gur* and Nanduben Patel poured some milk over it. 'I had not eaten such delicious food since coming to jail', she wrote, 'but I must not show it'. In a letter to Saraladevi, from Yeravda, Sardar Patel wrote that Maniben was trying to look after Mridula in Jail as well as she could, by reducing the inconveniences and also looking after her health, so she should not worry too much.[8]

On the day before her release from Belgaum jail, Mridula wrote in her diary:

Today is the last day that I have to wash my jail *saree*. In the whole ward, our *sarees* are the cleanest. We have to leave these *sarees* for the next lot of prisoners.... The Gujarat group all sat together today. Lilavati Munshi invited us in the 'B' ward for dinner; she put two flower vases and we felt as if we were having a picnic. I must confess that my heart was full and I could not control my tears. Since morning I have been unhappy that today is my last day. I loved the peaceful life here, no worries, no tension, no noise. I have learnt so much during these six months.[9]

She was released on 22nd June 1932. These six months were indeed an important phase of her life. In those days there was a fairly rigid 'caste system' among the owners of the Ahmedabad textile industry. Socially, the mill-owners (*shethias*) had hardly any contacts outside their own group. Mridula met in jail persons whom she otherwise would never have had an opportunity to associate with. Having grown up in a wealthy family, she had no idea of the daily problems of ordinary people. Jail life loosened the rigid social boundaries, widened her horizon and helped to shape her new life. Participation in the *satyagrahas* in the 1930s and her experience in jail profoundly influenced her personality. As Jawaharlal said, 'if we can succeed in prison in learning forbearance and the art of getting on with others who happen to be different from us, then we have learnt the secret of success in public activity, and indeed life itself'.[10]

Her jail experience was no doubt at least partially responsible for 'her de–classed thinking...all her co-workers had the opportunity of sharing the same dining table with her'.[11] She would happily sit in a Harijan colony and eat *khichdi* or drink tea in the houses of poor Muslims. Her friend Kapilrai Dave recalls how in their younger days when Mridula and he went to the cinema, she insisted on sitting in the cheapest seats, the so called 'pit' class with the mill-workers. The manager of the cinema hall, recognising her, would request her to sit in the balcony but she always refused to do so.

She was released but the *satyagraha* was going on and she could not keep away for long. It was as if she was itching to go back to jail. On 23rd January 1933, sub-inspector Bapalal of the secret service served a notice on her which was signed by the district magistrate of Ahmedabad prohibiting her from entering the walled city and Maskati Market for one month, or taking part in the Civil Disobedience Movement. She disobeyed the order and on 17th February, she was arrested again and produced before the district magistrate and kept in Sabarmati jail till 3rd March. Thus began her third experience of jail life. After being detained for some weeks in Sabarmati jail, she was shifted to Belgaum where the home secretary to the government of Bombay, N.W. Maxwell, issued an order directing her to reside and remain within the limits of Belgaum city and not to leave the limits demarcated without permission of the district magistrate. This order was to be in force for one month or till any other order was issued. During this period, her family members visited her in Belgaum and spent some time with her.

As was to be expected, she violated this order by picketing outside a cloth shop in Belgaum and was immediately arrested and sentenced to six months' simple imprisonment and fined Rs 500. Mridula was very worried as her brother, Suhrid and sister, Bharati were going to Oxford to study. The entire family was going to Europe and her parents were very keen that she should accompany them. Mridula disapproved of Suhrid and Bharati's going to Oxford. When a fellow prisoner asked her if she was happy about it, she said 'yes' and then regretted telling a lie. Gandhiji blessed Suhrid and Bharati's going. Mridula was most upset by this: 'I never understand why Bapu does this: says what people wish him to say. He never gives his true opinion. I do not approve of this at all'. She knew that her disapproval and her refusal to accompany her family on their trip to Europe was causing her parents great pain:

I will be released on 16th September, ten days after the family sails for Europe. But I do not want a remission of my sentence and an early release which would enable me to accompany them. From 1928 onwards, I have refused to accompany my family on holidays to hill stations or on trips to Europe. I know how disappointed Papa will be but I have to be firm.[12]

In jail, Mridula realized how deeply attached she was to her parents, 'I feel at peace and rest when Papa and Mummy are near. That they are near, even that knowledge gives me peace of mind. I have never shown this outwardly. If they are not there, I don't like going home'. She started embroidering something for her father and thinking of him, she did not get bored doing it:

Papa will get something, looking at which he will always think of me. He will like it because of the sentiments which have gone into it. I don't know why but today I am thinking of Papa and feel like crying only if I had some privacy but this is, of course, impossible. Many think I have no affection for Papa and Mummy but it is only because I don't know how to show my emotions. I don't want to do this either. Affection should be silent and demonstrated at the right time. When it will be tested, no one can say. Because of the struggle, I have started being away from home. At present my younger brothers and sisters are with our parents giving them company and joy, bringing life to the home. But when they leave, then it will be my turn to serve my parents.

To wrench herself from so much affection must have been difficult. She was constantly torn between her love for her parents and what she regarded as her duty. She knew that her involvement in politics and imprisonment was causing her parents a great deal of worry, concern and pain. She found this conflict at times unbearable: 'Will my whole life be like this? I would not mind if this affected me alone. But I do not want that for my sake Papa and Mummy's life should be sad or full of worries. May God give me the strength to face this test'.[13]

She eagerly waited for letters from home, which were often delayed because Gujarati letters took longer in censoring. The letters were often long—once it was seventy pages, in two parts! Sometimes she saw them in the jailor's pocket and recognized the handwriting but could do nothing. This taught her self-control and self-restraint.

She used to wait eagerly for the visitor's day when her parents would arrive. For two days before the meeting on 6 September 1933, she kept thinking of her home and visions of the Retreat floated before her eyes. She was terribly homesick and could almost see the beautiful garden, full of flowers of different colours and fragrance, flooded by moonlight, with her father standing in one corner. This meeting was going to be particularly painful because it was on the eve of Suhrid and Bharati's departure for Oxford. She kept thinking, 'Papa must now be doing this, now that, now they must have left Ahmedabad. The family must be having lunch. This will be Suhrid's and Bharati's last lunch at home before they leave for Oxford. When will Papa have so much joy again?.... But I must keep myself under control.' On the day of the visit, she broke down. She had tears in her eyes and was sniffing but pretended to her fellow prisoners that she had a cold and they believed her because few suspected her of being sentimental and prone to crying.[14]

On the day of the departure she thought of the family on their way to England; all of them had gone except her. She prayed for them at 12 o'clock, when the boat was to sail. 'Now it is 2.30 a.m... They must be out into the ocean and cannot see the coast of India. The ship is going farther and farther away'. Tears came to her eyes but she checked them. 'Let not your heart be troubled neither let it be afraid'—she consoled herself with lines such as these or by humming a verse from a Meera *bhajan* or a Gujarati prayer song.[15]

The family also missed her and Ambalal wrote: 'Just as there is an empty berth for you, so also there is space for you in our hearts and life. If you had been with us, father's pride would have been completed'.[16]

In Innsbruck, when the hotel porter told him that a telephone call from Mridula was expected at 1.25 p.m., he wanted to dance with joy. 'That I wanted to express my joy by dancing must be due to the air of Europe.'[17]

It was not always easy to put up with jail life. Mridula never complained but the physical discomforts were difficult to get used to. The weekly cleaning of the cell was a nuisance to one used to servants. Her fellow prisoners began sweeping and swabbing the floors at 6.30 in the morning. She had to get used to the noise and commotion and learn to sleep through it. Apart from the physical discomforts, there were times when she felt restless and depressed; in moments of strain and agitation the deprivations and loneliness of prison environment could become unbearable. The whole of August 1933 was a long, bad month of sleepless nights, worries, lack of peace. She suffered from constant headaches.[18] She shed more tears alone than ever before. *Christ in Silence* had a deep impact on her and lifted her out of her depression and loneliness:

For years I have had an idea of a true friend. But such friendship is not possible in real life. One can have colleagues, acquaintances but real genuine friends are very rare. I dare not call people friends in order to protect myself, prevent myself from being hurt.[19]

There was a further problem that her men friends were her fellow Congress workers like Purtan Buch, Indravadan Thakore or Kapilrai Dave; they came from a social class below hers and her parents did not particularly take to them. They too felt they should not become too intimate with a millionaire's daughter.

In prison, she reflected on the institution of marriage and how it was often only a means of begetting children or for preserving property and money. After marriage, a man started regarding his wife as his property and the wife also accepted this, an attitude deeply ingrained in women. She was opposed to this kind of marriage.[20] She could not conceive of such a marriage for herself. She was afraid of losing her independence, and feared that marriage may lead to bondage and responsibilities she was not yet prepared to shoulder:

I do not have the courage to enter the relationship of marriage. This is one matter in which my courage fails me. And it is good for I believe that without marriage, one can still have a platonic friendship with persons of the opposite sex. I am not interested in marriage for anything else but this companionship.

She felt that she did not want to waste her life in her youth on marriage. When she was older and wanted a companion and found a suitable one, then she could perhaps think about it.[21]

She was also influenced by Gandhiji's ideas on marriage. She read his *Experiments with Truth* in jail and noted in her diary that 'Gandhiji thought it impossible to combine service to society with family life. Those who really wanted to serve society, should observe *brahmacharya* as public life demanded full-time work and there were no half-measures. Only an unattached free person could do this[22]. Many of Gandhiji's women followers remained unmarried and he would have liked to compel them to take a vow of life-long chastity.

Mridula was eventually released on 26 September. Her next imprisonment was in December 1938 when a *satyagraha* was launched in Rajkot by the *Praja Parishad*. After the arrest of Maniben Patel, Mridula rushed to Rajkot to assume leadership. Her mother's family originally came from Rajkot and she had spent a part of her childhood there. When the people of Rajkot started demanding responsible government and launched a *satyagraha* for this, she was eager to take part in it. Bhaktilakshmi Desai and Kasturba were also active in Rajkot. Morarji Desai wrote to Mridula warning her that she would be arrested as soon as she reached Rajkot and asked her to look after her health. He asked her not to worry too much about the Congress, to have faith and spend her 'holiday' in joy and introspection.[23] Mridula was arrested as a dangerous person within forty-eight hours of her arrival in Rajkot and was sentenced to five weeks' imprisonment. She and Maniben were kept in Tramba Darbar's bungalow, three or four miles from Rajkot.

After her arrest in Rajkot, Subhas Bose in a telegram congratulated her and said that people were proud of her.[24] Nehru in a letter wrote 'My dear Mridu, I envy your quiet in prison and am glad that after your hard work you are having some peace. It was of course not necessary for you to ask for my good wishes in your new adventure. You have them, as always and my love, without asking for them.'[25]

Her parents, brothers and sisters visited her in jail. She was not keeping well and developed some throat trouble. Ambalal Sarabhai who was constantly concerned about her health, wrote to the jail superintendent requesting that their family physician, Dr. A.N. Tankaria be allowed to meet the jail doctor to discuss the treatment.[26]

The 'Quit India' resolution was passed on 8th August 1942 and in the early hours of the next day, Gandhiji and other leaders of the Congress were arrested. The movement which followed was, therefore, leaderless, and the younger and more militant cadres were left to their own initiatives. As the news of these arrests spread, spontaneous demonstrations broke out and women were quick to respond. As in the 1930s, they took out processions, held meetings, organized strikes in schools and colleges and distributed Congress leaflets and *patrikas*. Young girls and old women in different parts of Gujarat showed remarkable courage. Mridula was arrested on 20th August and taken to Arthur Road Jail, Bombay. She has vividly described the period of her detention:

I was arrested at Bombay on 20th August 1942. The warrant of arrest was issued by the authorities of Ahmedabad. It took more than ten days to arrest me. Soon after my arrest, I was lodged in the Arthur Road Jail, Bombay. Thereafter, I was transferred to Yeravada Jail, Poona and finally to Belgaum Jail. Nobody was allowed to meet me for the first six or eight months. Food was served to me not from the common kitchen of the prison but from outside because of the recommendation of the medical officer. I was categorised as 'B' Class prisoner. The most heartening thing for me was that Maniben Patel, daughter of Sardar Vallabhbhai Patel, was my companion. Shri Mahadev Desai had passed away in the Aga Khan Palace Jail on 18th August 1942. The British Government had suppressed the facts that led to his death. I was transferred from Yeravada Jail to Belgaum Jail. The police party which escorted me from Yeravada Jail to Belgaum Jail had come from the Aga Khan Palace Jail where Mahatma Gandhi was detained and from the Ahmed Nagar Jail where the members of the Congress Working Committee were imprisoned. Through these police personnel, I came to know about the happenings in the Aga Khan Palace Jail and Ahmed Nagar Jail. While being escorted from Yeravada Jail to Belgaum Jail, I gave news to passers-by about the details of the death of Shri Mahadev Desai and the future course of action which the leaders contemplated to take. Thus, a large number of people came to know what had happened in the Aga Khan Palace Jail and the Ahmed Nagar Fort.

The wives of many leaders were with us at the Yeravada Jail during this period of detention. The male and female prisoners were lodged in two separate buildings which were adjacent to each other. A common road divided them. Permission had been given by the Jail Authorities for the joint preparation of food for 'A' class and 'B' class prisoners. On arrival of female prisoners in the Jail, the male prisoners asked them to prepare food for them. The female prisoners retorted that they had been cooking for them throughout their life; could not they cook for them [for a change?] The result was that the male members took to cooking and within no time became experts....

When Mridula was in Belgaum Jail, cholera broke out in the jail. The disease assumed epidemic proportions and the jail authorities were seriously concerned. Mridula did a great deal to look after the cholera patients, to build up their morale and entertain them. When there was a move to transfer some of the important prisoners, Mridula pretended to be ill and, therefore was not removed. The Government wanted to reduce the number of prisoners because of cholera and so allowed them to be released on parole. But none of the prisoners was prepared to be so released because if they were released on parole, it was obligatory to give in writing that they would not do anything in future which would be against the government.

On 13 September 1942, Mridula's three sisters Bharati, Gita and Gira had also been arrested for having taken part in a procession. Each of them was sentenced to three months' imprisonment and fined Rs 200. Two of their aunts Indumati Chimanlal and Nirmalaben Bakubhai were also arrested. Thus at one time five women of the Sarabhai family were behind the bars, in addition to Mridula. During this time, Suhrid fell seriously ill and Ambalal requested H.V.R. Iyengar, the then Home Secretary of the Bombay Presidency, to release them. The Governor, Sir Roger Lumley, with whom Iyengar took up the matter agreed to release them immediately on parole through telephonic orders.[27] While her sisters, Bharati, Gita and Gira and her aunts agreed to be released on parole, Mridula refused to accept parole on principle and demanded unconditional release. This was denied and, therefore, she could not see her dying brother. Ambalal wrote that he appreciated her difficulty in giving an undertaking and did not misunderstand her not coming out on parole due to certain vital principles and the responsibility she felt for her work.[28] In fact, when the telegram informing her of Suhrid's death reached her, the prisoners were celebrating the birth of a new baby in jail. Mridula had specially ordered some sweets for the occasion. When the warden brought the telegram and asked her what she should do about the celebration in the evening, Mridula said that everything was to go on as planned. She hid the telegram in order not to spoil the celebration and bore her sorrow in secret. That night, Mrinalini Desai, a fellow prisoner, on waking up suddenly, did not find Mridula beside her on her bedding on the floor. Getting up, she found her sitting on the floor, holding the bars of their cell door weeping silently. She held Mrinalini's hand and both of them cried. When Maniben Patel woke up, Mridula embraced her and burst into tears saying 'Maniben, Suhrid has gone.[29]

On Suhrid's first death anniversary, Mridula wrote a very touching and affectionate letter to her parents from prison asking them to follow Suhrid's example and to face his loss bravely. She felt that instead of trying to unravel the mysteries of life and death, raising questions to which perhaps there were no answers, they should try to involve themselves in improving things around them.[30]

Mridula's health deteriorated in jail but to seek release on grounds of illness was unthinkable for a *satyagrahi;* even to lodge a complaint with the jail authorities was not proper. She, therefore, did not tell anybody about her illness. Maniben Patel who was aware of this drew it to the attention of the jail superintendent. As a result, Mridula was taken to J.J. Hospital, Bombay. The doctors examined her and decided that she should be operated upon for appendicitis.

'I was told by the British authorities that I could be released on parole provided I don't meet any Congressmen. I did not agree to this condition. I retorted that I do not have acquaintance with anybody except Congressmen. I was released unconditionally in December 1943 on the day when I was to be operated upon.' This was her last imprisonment before independence.

CHAPTER FOUR

THE INDIAN NATIONAL CONGRESS

The Indian freedom struggle under the direction of the Indian National Congress captured the imagination of and inspired people all over Gujarat. The Gujarat Congress had undertaken relief work under the leadership of Sardar Vallabhbhai Patel when bubonic plague broke out in Ahmedabad in 1917. Vallabhbhai was chairman of the Sanitary Committee of the municipality and for three months he and his volunteers worked day and night to bring the epidemic under control. The Kheda *satyagraha* of 1918 saw Vallabhbhai emerge as a hero. Ten years later, Ahmedabad, indeed the whole of Gujarat and Saurashtra, were stricken with another calamity—torrential rains and floods. Vallabhbhai lost no time in organizing relief parties to provide immediate shelter and food to the victims. He moved with remarkable speed and efficiency to bring relief to the flood-stricken areas.[1] As a result of this, he came to be known as *Gujarat Vallabh* (the glory of Gujarat), became immensely popular and won his place as *Sardar* in the hearts of the people of Gujarat.[2] The Bardoli *satyagraha* was equally impressive and confirmed the popular notion that the Congress was an organization capable of delivering the goods. When the Congress held its annual session in Ahmedabad in 1921, there was tremendous enthusiasm and jubilation in the city. Young Mridula, then only ten years old, went to the session with her parents and set up a stall to offer water to the delegates. She joined the *Bal Sena* (Children's Brigade) of the Congress and the *Shishu Vibhag* (children's section) of the *Akhil Bharat Charkha Sangh*. (All India Spinner's Association).

Mridula became a member of the Congress in 1930 and was chosen as a delegate from Ahmedabad for it's Lucknow Session in 1936, over which Jawaharlal Nehru presided. She was made responsible for setting up a Women's Department in the Gujarat Provincial Congress and was appointed its Joint Secretary. In 1936, the Gujarat

Provincial Congress Committee (GPCC) formed a committee, with Mridula as convenor, to examine how women's representation in the All India Congress Committee (AICC) could be increased and to collect information regarding women's organizations and their social and economic position. The object was to draw women who had taken part in the Civil Disobedience Movement, into public life again. Mridula was elected a member of the AICC from Gujarat in 1936–7 and took up the task of mobilizing women voters for the provincial elections held in 1937.

The Haripura Congress of 1938 was a great experience for Mridula. She was appointed captain of the *swayam sevikas* (women volunteers) by Sardar Patel and was in charge of their recruitment and training, while Harivadan Thakore was in charge of the men volunteers. They worked in close co-operation under the guidance of Morarji Desai who was in overall command of the Congress Sevadal. She travelled all over Gujarat, 'it was the most interesting, instructive and enjoyable tour that I ever had',[3] she wrote thirtyfour years later. A training camp for the volunteers was set up and as soon as the first group of women were trained, they were sent back to their respective districts to initiate their own training programmes.

About five weeks before the Congress session, nearly two thousand women volunteers reached Haripura. Vanmala Desai, one of the women workers selected for recruiting and training volunteers, had to undergo training at Bardoli herself. She was then put in charge of training two hundred women volunteers. She recalls how at Haripura, Mridulaben was available at night till 12 and was up again at 4 a.m. Young volunteers had free access to her day and night and she looked after every minute detail. 'She lived in a hut like us, ate the same food and shared our discomforts. She was strict but at the same time kind'.[4]

Harivadan Thakore who was in charge of the men volunteers recalls her amazing organizing ability, patience, perseverance and capacity for ceaseless work at Haripura.[5]

Volunteers had to be recruited for the foundation stone laying ceremony. Mridula and her colleagues toured villages within a radius of ten miles of Ahmedabad and in seven days were able to raise men and women volunteers who were given the necessary training. On the opening day of the Haripura Congress Sardar Patel and the other leaders were delayed in arriving for the salutation. Due to the heat and exhaustion just when the singing of *Vande Mataram* started,

Mridula and a few other volunteers fainted. On the closing day of Haripura Congress, during the flag salutation ceremony, Sardar Patel stood behind Mridula and Harivadan Thakore, with his hands on their shoulders while *Vande Mataram* was being sung. Afterwards, Sardar heaved a sigh of relief and said, 'I stood there to see that my volunteer captains do not collapse during this function.[6] Mridula was deeply touched by this.

The women volunteers who had come to Haripura became the nucleus for contact with the women of their respective villages and towns. Through them it was easy to get the message of the Congress Women's Wing to Gujarat and Saurashtra at minimum cost and effort. It is often argued that women who took part in the Gandhian movement did not really change their role within the family and the home. Mridula's experience was different. Many of the men who had sent their wives, sisters or daughters wrote appreciative letters to Mridula telling her how these women had become more self-confident and self-reliant as a result of their training. A number of men wrote that they were grateful to her for having revolutionized their family life, allowing them to live as comrades, equally sharing the happiness and stresses of family life.

With the success of the volunteer programme at Haripura, the GPCC President, Morarji Desai, decided to form a permanent volunteer corps at the *taluka* and village level. The training was to be modelled on the Haripura camp and the women's department was placed under Mridula.

While Mridula regarded the Congress as the only organization which could achieve the country's political freedom and also bring about the desired economic and social changes, she had differences of opinion with many Congressmen. One of the issues was their attitude towards women. In 1934, in Gujarat, for election purposes, women were put in the same category as backward classes by the Congress. She resented this; after their participation in the Civil Disobedience Movement, how could they be categorized as backward? She protested against the non-inclusion of women in the Congress Working Committee in 1936, by writing letters to Jawaharlal, Gandhiji and others and made the women in the Gujarat Congress also write a letter of protest. To Gandhiji, she wrote:

Congressmen are not interested in understanding the attitudes and views of women who are taking part in public life. In politics, women stand by the Congress, but it is not sympathetic to their social and economic problems.

We are loyal Congress women and feel that the Congress is the only organization through which our women's problems can be solved.[7]

She wanted the Congress to take a positive stand on women's rights and fight social evils and injustice against women. To Vallabhbhai, she complained that the Congress opposes those women who are divorced or re-married or women who marry men who have first wives, but no objection is raised against the men who do so. In Haripura, some Congressmen objected to certain women volunteers on the ground that they did not have a good reputation. Mridula replied that she and her colleagues did not approve of some of the male office bearers of the Reception Committee. Would they be removed?[8] Under similar circumstances a common standard should be applied to both men and women. She also had some difficulty with her male colleagues because while she never felt that as a woman she should be treated differently, they, contrarily, could not treat her as an equal. She hoped that as the number of women workers in the Congress would increase, this attitude would change.

Mridula organized a women's department within the GPCC and initiated moves for the formation of a women's department in the AICC. It took her a long time to convince Congressmen of the necessity for such a department and even when it was started, its progress was extremely slow. The response from the provincial and local Congress committees was lukewarm. Her complaint was that most Congressmen were male chauvinists and not really interested in women's problems. In 1946, she wrote to Vallabhbhai opposing reserved seats for women in the new constitution that was being framed, as she wanted women to be treated at par with men, as equals.[9]

The undisputed leader of the Gujarat Provincial Congress Committee since its formation in 1920 was Sardar Patel who was also its first president, a position he retained till 1946. He was a trusted lieutenant of the Mahatma, with a growing reputation as a 'strong' man. His organizational skills and his toughness of fibre were recognized by his friends no less than his enemies. Vallabhbhai himself obeyed Gandhiji; in fact, 'when it came to obeying an order there were few like him.' He expected the same loyalty and obedience from his partymen. Within the Gujarat Congress, hardly anyone dared cross his path or even speak out his mind freely if he disagreed with him. Mridula was used to independence of thought and freedom of expression. She never hesitated to express herself freely in private as well as in party meetings when she differed from the Sardar or his lieutenants. He made no secret of his disapproval of such independence and openly frowned upon her nonconformity. But she went her own way at the cost of her own political future in Gujarat where she was never quite regarded as 'one of us' by those in command. Her father wrote to her,

It is a great pity the local Congress organisation should not take advantage of your capacity, zeal and sincerity. But I despair of any improvement so long as Sardar controls the work in Gujarat. With all his good qualities he is very jealous of power, intolerant of differing views and prefers that both quality and quantity of the Congress work suffer with 'yes' men to improving the quality of work and getting capable men who will not on all occasions say 'yes'.[10]

Yet the Sarabhai family's relations with Vallabhbhai, as those of most Ahmedabad industrialists, were close and he had a great deal of personal affection for Mridula and the entire family. Ambalal wrote to Bharati '... Vallabhbhai was here and he dined with us on Saturday night.... There was a pleasant and interesting conversation during dinner and afterwards. Vallabhbhai appeared to be overworked, but he was, as before, very clear in thinking and very witty....'[11] Vallabhbhai's letters to Mridula when they were both in jail, often addressing her as 'Mridu', show his deep concern for her health and well-being. When he was in jail in 1931 and she was outside, he asked her to write to him every week, giving him all the family as well as political news. He could receive four letters a week of which he wanted one from her.[12] He wrote to her regularly in the years 1930 to 1934. In every letter he enquired about her parents, brothers and sisters. Once he came to Ahmedabad and the Sarabhais were all away. 'I spent a week in Ahmedabad and remembered all of you. I missed all of you', he wrote.[13] In 1947, when Mridula was in Amritsar and her parents in Ceylon, Sardar telephoned them one night to tell them that she was safe and well.[14] Mridula owed her induction into public life in a way to him. 'In public life, the kindness and sympathy that you have shown me', she wrote to Vallabhbhai, 'I shall always remember and try and prove worthy of it'.[15]

Though for years she had been close to the Sardar and Maniben, Mridula later developed a closer association with Jawaharlal. On questions such as socialism and the Hindu-Muslim tension she and Jawaharlal were on the same wavelength which was not Vallabhbhai's. Her attitude, which they regarded as a switch in loyalty, hurt both Patel and Maniben. The Sardar wrote to her:

I knew that you had changed your views. But I purposely never asked you....
But you should not accuse the Working Committee of dishonesty. Even though
there are differences of views, you should not doubt the honesty of your
opponents, that is the first lesson of public life....[16]

Mridula hastened to clarify, rather surprisingly: 'I know that you will
not misunderstand me expressing my views and I believe that you
have always given people the freedom to express their views even if
they differ from yours'.[17] Unusual for her, she was attempting to be
conciliatory, but evidently did not succeed in pleasing Sardar.

She found many of Patel's lieutenants in Gujarat, too rigid, too
conservative and too cautious for her liking. Her method of working
was also quite different from theirs. She was a doer with a positive
approach but these men, she felt, were negative, always raising objections to any new scheme or plan she proposed.

Her relations with Morarji Desai were close specially in the years
before independence. Despite differences over many matters, they had
personal affection for each other and a sense that they were fighting
for a common cause—the country's freedom. Morarji knew the family
well; in 1931 at the Karachi session, he stayed with the Sarabhais.[18]
When they were in jail, they recommended and sent books to each
other. Morarjibhai admired her capacity for hard work: 'You will have
to follow up the Report by action and you alone can do it', he wrote,
'because we have very few workers with your enthusiasm and
capacity for thoroughness, going into minute details.[19] They spoke
and wrote to each other frankly. Morarjibhai, on more than one occasion, wrote that as long as their differences were honest, and they
did not doubt each other's motives and intentions, he did not mind
their disagreeing at all. Once, when she accused him of sarcasm, he
replied that he knew that he had a reputation for saying things
straight, unpleasant and bitter but not sarcastic.[20]

The Gujarat Congress leaders were against holding a Youth Congress in Gujarat which was led by Jawaharlal Nehru and Subhas Bose,
two rivals for capturing the leadership of the youth. Sardar disap-
proved of any kind of populism and looked down on the Youth
League which according to him indulged in rhetoric and no action.
In 1928, against the wishes of the Gujarat Congress leaders, a Youth
Conference was held at Rajkot under Jawaharlal's presidentship;
Mridula was one of the organizers. In 1936, she came into open con-
flict with the GPCC on the question of having voted in favour of
Nehru's resolution for non-participation in the parliamentary

programme at the Lucknow Congress. The Gujarat Congress wanted to take disciplinary action against her for flouting their directions and voting against their instructions for breaking the age-old tradition of the GPCC that delegates to the AICC would always vote, according to instructions, in a block. But Mridula was not prepared to bind herself in any way and protested against this convention. She said that the procedure of previous discussion and mutual consultation with delegates had not been followed. The Gujarat Congress wanted her to resign from the AICC delegateship. According to them, she had got this position on their having included her in their panel and so she was duty bound to resign. She offered to resign but made it clear that she would contest next time as an independent candidate. Gandhiji intervened and the matter was dropped for the time being.

Vallabhbhai put her on the Executive of the GPCC, but did not support her election as a delegate to the AICC. To Jawaharlal, she wrote: 'Apparently, Gujarat looks to be united; but as a matter of fact that solidarity has been in a process of disintegration. Discontent is secretly smouldering. The Congress needs to be strengthened here. Also it is necessary that Gujarat should come into your direct contact. In the last Muncipal election of the city, the local Congress Committee had put thirtyfive candidates on the Congress ticket; only twentyone of them have succeeded... This landslide has opened the eyes of the local official Congressites'.[21]

Delegates to the AICC were selected on the basis of personal likes and dislikes, and the socialists were kept out. Mridula was sarcastic about Gujarat Congressmen professing to be followers of *satya*, and *ahimsa*. She contested as an independent candidate without official support but topped the poll.

Mridula was close to Congress socialist leaders like Jayaprakash Narayan, Ashok Mehta, Achyut Patwardhan and Ram Manohar Lohia. Whenever they came to Ahmedabad she looked after them. When Jawaharlal invited her to a meeting of the Congress Working Committee in Delhi in March 1937, she complained about the attitude of the Gujarat Congress, particularly towards the socialists. Vallabhbhai conveyed her complaints to Morarji Desai, who in turn wrote to Mridula in a reprimanding tone. The Gujarat Congress once again did not select her as a candidate to the AICC in 1940, so she asked Jawaharlal whether she should stand without their support as an independent candidate. Nehru, as was often his manner, gave an ambivalent, non-committal reply: 'There is no harm in your standing for elections and getting defeated. On the other hand, if you do not feel like it, then there is no point in standing'.[22]

Not only did Mridula regard many of the Gujarat Congress leaders as reactionary but also as communal. When Hindu-Muslim riots broke out in Ahmedabad in 1941, she found the Congress House most apathetic and passive. She took command of the situation and directed relief operations and undertook a vigorous campaign for restoring communal peace. She was one of the secretaries of the Ahmedabad Congress from 25th May to 10th October 1941, as Congress leaders including Vallabhbhai, Morarji and some others were in jail because of the 'individual *satyagraha*'. Mridula wrote long letters to them as well as to Gandhiji, drawing their attention to the lack of leadership and pro-Hindu communal bias shown by the Gujarat Congress. She presented a detailed report of the work done during these months. The two Gujarat Congress leaders with whom she had special rapport were Gulzarilal Nanda and Khandubhai Desai of the Majoor Mahajan, both of whom she found to be non-communal.

In the 1969 riots, she was similarly active but the Gujarat Congress leadership suspected her of being pro-Muslim and an agent of Sheikh Abdullah and prevented her from working in Muslim refugee camps.

Whenever Mridula felt that the Congress was doing something she did not consider proper or right, she unhesitatingly wrote to Gandhiji, Jawaharlal, Vallabhbhai or Morarjibhai saying so. In 1935, she expressed her disapproval of the Congress spending money on illuminations and fireworks on its golden jubilee celebrations. In the 1946 elections when the Bombay Provincial Congress Committee selected Lathe as a candidate, she expressed her disapproval in letters to Sardar and Morarji.[23] She felt that the Congress was not valuing those who had been loyal and disciplined followers and was instead favouring those who had not done a day's service in the cause of freedom and had been pro-British. On these grounds, she opposed Congress support to even Rajaji and Chintamani Deshmukh. Vallabhbhai replied, 'We need different types of persons for legislatures. You must be broad-minded in public life'.[24] In the same manner, she objected to the U.P. Congress selecting candidates whom she regarded as definitely communal.

The results of the Gujarat Provincial Congress Committee elections in 1946 were unexpected and showed resentment against the dominant group. There were keen contests in Surat, Ahmedabad, Palanpur and in two or three other places. In many cases, the official candidates were defeated. Mridula was elected as an independent. To Jairamdas Daulatram she wrote: 'I stood as an independent candidate as I have been doing all these years. My position is a bit awkward you know. I never belong to the official group here, and naturally they do not like it....'[25]

In July 1946, Nehru, who-was the president of the Congress in that year, nominated her to the Working Committee. After Sardar, she was the first Gujarati to occupy this position. She was also appointed, together with B.V. Keskar, general secretary of the AICC, the only woman so far to have held this office. To Sadiq Ali, office secretary of AICC she wrote,

My appointment as the General Secretary of the AICC has been for me so unexpected and undreamt of. I came to know of it for the first time when the official announcement was going to the Press. Now that this has happened, I consider it an honour but fully realize also the responsibility and gravity of the appointment.[26]

As the Congress head office was in Allahabad, she moved there and stayed at Anand Bhavan. From there she wrote to her family;

Jawaharlal is a hard task master but if you satisfy his demands, the appreciation, comradeship and understanding you get from him, you can't find from others. He admires resourcefulness and efficiency. As long as I have Jawaharlal's backing I am prepared to face all opposition.[27]

She stayed in Anand Bhavan at Nehru's request and occupied Krishna's (i.e. Nehru's sister) room. Gandhiji wrote in *Harijan*

It is a good thing a woman fills for the first time in the history of the Congress the post of General Secretary. Shrimati Mridula Sarabhai was one of the pupils of Acharya Kripalani in the initial stage of the career of Gujarat Vidyapith. Therefore, she will have full guidance from her Acharya in the difficult task to which she is called.[28]

In that year, the GPCC formed a sub-commitee comprising Morarjibhai, Bhogilal Lala, Khandubhai Desai and Mridula to suggest changes in the Congress Constitution. During her short tenure as AICC secretary, she tried to reorganize and vitalize the Congress office. She held that the office was capable of great expansion and could undertake many new useful activities.

She was secretary, however, only for four months as in November 1946, the Congress joined the Interim Government and Nehru became the Prime Minister. He, thereupon, resigned as Congress President and Acharya Kripalani succeeded him. By this time differences between

Nehru and Kripalani had started developing. The latter regarded Mridula as too close to the former and also held that she was not an orthodox Gandhian. Therefore, he did not re-appoint her to the Working Committee. According to Eleanor Morton, of all Gandhiji's women disciples, Mridula saw him with the clearest eyes. She said to him firmly "I am no Gandhian, remember!' It was the work, not the leader, which held her allegiance. He loved to hear her saying this; his laugher would ring out in appreciations".

To Kripalani she wrote:

I have always felt that I have been misunderstood. As you already know from 1920 to 1928, even though young and in an impressionable age, I tried to train myself in the 'Gandhian way'. I had a great regard for the 'Gandhians' and would have nearly associated myself with them, but for the fact that from 1928 I started coming into direct contact with them. Experience in college and public life opened my eyes. For the last 15 years this controversy has come in the way of colleagues entrusting me with any responsible work. I do not regret this, because it has allowed me to remain honest to myself....[30]

Kripalani said that she sought her light and inspiration from two sources—Gandhiji and Socialism. 'It cost me many anxious hours to decide the question of your remaining general secretary. You know that even when we were very friendly, we did not see eye to eye in politics, and therefore, would not be able to pull on together'.[31]

Mridula was critical of the way the Navjivan Karyalaya and Gujarat Vidyapeeth were functioning. She complained to Gandhiji about this. She felt that Gandhians had a smug, complacent attitude and were unwilling to accept any criticism. Gandhiji conveyed her complaints to Jivanji Desai.[32] Yet she was very close to Gandhiji. He had written to her in 1934, 'I will not stop guiding you or advising you because I expect great things from you.... My blessings are always with you'.[33] And again, 'I would like you to come every three months to see me. More often if possible'.[34] In September 1944, she together with Khurshedben was on Gandhiji's office staff. 'They are all working full speed—not to mention Pyarelal, Sushila and Kanu', wrote Bapu to Miraben.[35]

The fiftyfirst session of the Congress in 1946 at Meerut was a historic one as it was the last session before Independence, where the Congress accepted the Cabinet Mission Plan and agreed to a Constituent Assembly. Just before the meeting, communal riots broke out in Meerut and Mridula rushed there from Delhi to help in restoring peace. From Meerut she also had to drive to the village of Shah-jahanpur where another outbreak of communal violence was feared.

In 1950, a controversy developed over the leadership of the Congress, somewhat similar in nature to that which had arisen at Tripuri (a fancy name for a rustic site near Jabalpore in Madhya Pradesh) in March, 1939. Then, the candidature of Subhas Bose had been opposed first by Maulana Azad and then by Pattabhi Sitaramayya. Subhas Bose was elected but had to resign because of the opposition of Gandhiji and the Congress High Command. In 1950 there was a triangular contest between Purshottamdas Tandon, Shankarrao Deo and Acharya Kripalani. This was perhaps the last of the several 'conflicts' between Nehru and Patel. Patel threw his support behind Tandon who was said to be a Hindu communalist. Nehru supported first Deo and then Kripalani, as he regarded Tandon as a communal reactionary. Yet Tandon was highly respected for his integrity and many who did not share his views were prepared to vote for him. His election rival, J.B Kripalani, a devoted follower of Gandhiji since 1917 and a member of the Congress high command since the mid-thirties, had served for twelve years as general secretary and in the crucial year of partition as president of the Congress.

While fully supporting Tandon, Vallabhbhai was prepared to consider a candidate whom both he and Nehru could support. Maulana Azad proposed S.K. Patil, and when Vallabhbhai agreed, sought Nehru's approval. Jawaharlal rejected Patil's name and rumours were floated (Patel believed by Rafi Ahmed Kidwai and Mridula) that Nehru would become president.[36] Mridula opposed Tandon's candidature and tried to convince Kishorilal Mashruwala, Morarji Desai and Lal Bahadur Shastri among others of the inadvisability of electing Tandon as president. Mashruwala wrote to her that he had heard that Tandon was not a communalist and that the campaign against him was being conducted by Rafi Ahmed Kidwai, who was also poisoning Nehru's ears against Sardar Patel. An interview given by Nehru to one Ramachandran, in which the former had expressed his views against Tandon becoming President of the Congress, was leaked out, circulated and published in *Gujarat Samachar* on the day before the election. Some suspected Mridula's hand in this. Later, there was a denial from the prime minister's office of any such interview having been granted by Nehru at all. At the last minute, Shankarrao Deo withdrew from the race. The issue was fought out at the Nasik Session of the AICC in September 1950. Patel and the conservatives scored a victory with the election of Tandon, who secured 1,306 votes as against the 1,092 which went to Acharya Kripalani. Before the year ended, Sardar died and after months of friction, Tandon was forced to resign and Nehru took over as Congress President.

In the same year, a controversy arose between Mridula and the GPCC over her membership form which was lost in the local Congress Committee office and her name was struck off the rolls apparently because of some clerical error. As a result, she was not eligible to attend the forthcoming Congress Session as a member. Restoration of the right was essential, otherwise it would amount to break in continuation of membership, leading to disqualification from 'active membership' and the deprivation of her right to vote and stand for election. She was most upset and wrote letters to Kantilal Ghia, Secretary of the Ahmedabad Congress Committee, Morarji Desai and Pattabhi Sitarmayya, President of the Congress asking that she be restored as a member of the Congress.[37]

Mridula wanted to stand as a delegate to the AICC from Santrampur in the Panchmahal district of Gujarat in 1951 but the Congress candidate was not prepared to withdraw in her favour, so she finally gave up the idea and opted out of the contest.

Mridula was never keen on any office herself, nor on becoming a member of the State Assembly or Parliament, but she worked hard for Congress candidates whom she approved of and in the 1951–2 elections, mounted propaganda for the Congress in Gujarat and for Jawaharlal in Allahabad.

In 1953, when she took up the cause of Sheikh Abdullah, a campaign was started against her by some Congressmen who called her a 'traitor'. In April 1958, Sriman Narayan, Congress General Secretary, wrote a letter to the GPCC asking it to take disciplinary action against Mridula for her 'anti-national' activities as a journalist. Earlier, the Kashmir premier, Bakshi Ghulam Mohammed, at a public meeting in Delhi and at a press conference in Bombay, had accused her of carrying on anti-Kashmir government propaganda in the course of her campaign for Sheikh Abdullah[38] The GPCC sent her a letter asking her why disciplinary action should not be taken against her as her activities were 'against the declared policies of the Congress and detrimental to the interest of the country'.[39]

The Congress President, U.N. Dhebar, attended the Gujarat Provincial Congress Committee meeting on 28th June 1958 where it was decided to remove Mridula's name from the register of the Congress and debar her from being enrolled as a member for five years from the date of the resolution. The GPCC notes that:

Its executive had invited Kumari Mridula Ambalal Sarabhai to give her explanation regarding the notice issued to her. Instead of giving an explanation of her conduct, she had been putting counter questions, avoiding the explanation and had also evaded appearing before the Executive Committee of the GPCC.[40]

Mridula appealed to the AICC from detention in the Central Jail in Delhi against the disciplinary action taken against her by the GPCC, on the grounds that details of the charges against her were not furnished; only vague allegations were made and orders had been passed without a just and proper enquiry. She wanted a thorough enquiry before a final decision was taken. She wished to be judged, she said, by 'the yard-stick of the code of conduct for a national and emotional integration worker. The issuing of circulars and articles was no offence for a journalist'.[41] She had not 'revolted' against the Congress but was only drawing the attention of national leaders to their policy of trusting the wrong people in Kashmir. She was prepared to appear before the Executive Committee of the Congress if specific charges were framed against her. She was convinced that she had done nothing wrong in supporting the people of Kashmir and their popular leader and that through her work she had served the Union Government, the Congress and the country, and the Jammu and Kashmir State to the best of her ability.[42] After her release she submitted a supplementary note to her appeal from jail.[43]

The disciplinary action committee of the AICC called Mridula to meet them on 25th April 1960 in Morarjibhai's office in Parliament House, New Delhi. Besides Morarji, Balvantrai Mehta, Lal Bahadur Shastri and Jagjivan Ram were present. The interview lasted half-an-hour and she was not at all satisfied. Only Lal Bahadur Shastri and Jagjivan Ram were, she found, in a mood to listen to her and understand her position. Morarji said that as a Congress member she had no right to issue these circulars and 'that her activities were treasonable'. She was deeply hurt by his impatient and unfriendly attitude considering the years they had known each other and worked together. The Committee wanted her to give up her activities as a journalist and stop circulating anything concerning Kashmir. She said that she had broken no Congress Resolution on Kashmir and her circulars had not harmed India. She could not, therefore, accept their charges against her. Shastri told her, 'If a person like you and of your standing disagrees with the Congress, why do you insist on being in the Congress?' Mridula answered:

As far as the Congress is concerned, I have grown up in it and the question why we are in it or not does not arise. Since 1946 I have worked as a *Shanti Sainik* for the *Quami Ekta* Programme and the Congress had allowed this, but now you are refusing to accept me as a *Shanti Sainik* working for national and emotional integration'.

Later in the evening she went to meet Morarji to find out why he was in such a hostile and angry mood, but she sadly realized that the Congress no longer wanted her.

Thus ended her formal association with an organization with which she had wholly identified herself for over thirty years.

CHAPTER FIVE

TOWARDS WOMEN'S EQUALITY

In Society

Since her teens Mridula had protested against women being treated
as separate and inferior. Any form of injustice aroused her passion
and from a young age she felt angry at the inferior position women
occupied in the family and outside. When she was only seventeen
years old, at a Youth League Convention held in Ahmedabad, she
spoke spiritedly for removing all social distinctions between men and
women as well as between unmarried girls, married women and
widows. She objected to the attribution of different forms of address
for women, depending on whether they were married or not, and only
one for men. This discrepancy she argued, was because men's marital
status was deemed to be less crucial to their identity than women's.
Women's status had always depended on being attached to a man.
In Gujarati, an unmarried girl was addressed as *kumari,* a married
girl as *saubhagyavati* and a widow as *gangaswarup.* All women she
said should be addressed as *shrimati.* While married women wore
glass bangles, *mangalsutra, kumkum* and *chandla,* widows were
prohibited all these and had to wear only white *saris.* She protested
against this also and urged that there should be no difference in dress,
between unmarried girls, married women and widows. There was
considerable opposition from the audience to this and only two
women present at the Convention, Sharda Mehta and Pramoda
Gosalia supported her and took a pledge not to wear glass bangles,
mangalsutra and *chandla.*

The Youth League (*Yuvak Sangh*) which was quite active in Ahmedabad during those years, organized a youth week in 1929, during
which Mridula was responsible for arranging lectures, debates, discussions and cultural programmes. At one of its meetings, the
socialist leader, Rohit Mehta, moved a resolution that women should
not be addressed as *saubhagyawati, kumari* and *gangaswarup*; to

which Mridula moved an amendment that all women should be addressed as *shrimati* and widows should be permitted to wear coloured *saris,* jewellery, *chandla* and keep long hair. The resolution, together with the amendment, was passed by a majority. Rohit Mehta moved another resolution supporting widow re-marriage. Here also Mridula moved an amendment that widows should not only be permitted but positively encouraged to re-marry and steps should be taken to improve their position both within the home and in society. The resolution together with the amendments was adopted at the meeting.

Mridula and many of her young colleagues felt that women paid too much attention to their appearance, dress, jewellery and cosmetics which were supposed 'to make women cultured' but 'made them artificial'.[1] She tried to get rid of all those feminine qualities which she regarded as hindering women's progress, 'shyness, softness, helplessness, dependency, lady-like manners.... I don't want to be a doll for show either for men or society. I don't want to be artificial, unnatural.... I have tried to keep away from the so called womanly qualities and, tried to cultivate certain manly qualities which I consider essential for a woman—desire for adventure, daringness, self-confidence, discipline, ability to do one's work, control one's mind and emotions; one's physique and way of walking should be that of a soldier.'[2] She regarded herself as a soldier in the battle for freedom and deliberately cultivated qualities which would be commonly regarded as masculine.

Marie Seton refers to her as 'the mannish disciple of the Mahatma.'[3] The Yuvak Sangh arranged physical training classes for women and advocated simplicity in dress and giving up jewellery. In this they were influenced by Gandhiji who believed that wearing noserings, bangles, necklaces, anklets by women was in a way a symbol of their subordination to men; women should not use ornaments if they wanted to participate in the freedom struggle and work towards women's emancipation.

Women in large numbers had taken part in the Civil Disobedience Movement and Mridula was concerned lest the women who had been energized and awakened should relapse into apathy and go back to be confined to their homes. Gandhiji wrote to her:

Our women joined the salt *satyagraha.* They came out of their homes. It is now your duty to see that they should not be imprisoned within the four walls of their homes.[4]

To decide on what was to be done, a committee was formed with Shardaben Mehta as President. After long and often acrimonious deliberations a women's organization run by women for women, named Jyoti Sangh was established on 24th April, 1934: Gandhiji laid its foundation in the compound of Ambalal Sarabhai's bungalow Shanti Sadan, in the Mirzapur area of Ahmedabad. '*Jyoti* means light, just as a lamp gives light, may your efforts spread light into the houses of the poor'[5] was his message to the enthusiastic women assembled on the occasion. Jyoti Sangh, as its name suggests, was formed with the objective of giving light to women who needed it and to provide them with sufficient opportunities for their physical and mental development so that they could gain self-confidence and become self-reliant, and also contribute their due share in the development of the nation as enlightened citizens.[6]

Mridula was very keen on training women social workers and made what was perhaps the first such attempt in India through Jyoti Sangh. She herself taught in the department of training workers.

The work of Jyoti Sangh was divided into four main sections: Talimi Kendra, Adult Education, Udyog Vibhag and Sanchalan Vibhag. In the Talimi Kendra various vocational classes were held such as tailoring, embroidery, spinning, music, painting, drawing, etc. Mridula tried to get the most reputed teachers for each subject. Music teachers were brought from the Gandharva Mahavidyalaya and art teachers from the J.J. School of Arts, Bombay. The teachers were paid salaries and the students, stipends. The academic side of the programme included classes for teaching Hindi, Gujarati, arithmetic, current topics, civics and hygienec. A hostel was run for women who came from outside Ahmedabad. Summer vacation classes were conducted in various localities for school and college girls. To encourage women to read, books were borrowed from various libraries; for women who could not come to the Sangh, arrangements were made to deliver books at their homes. Lectures were delivered by distinguished politicians, social workers, economists, scientists, philosophers, poets and writers. Eminent persons passing through Ahmedabad were invited to deliver lectures. Those invited included Gandhiji, Sardar Patel, Jawaharlal Nehru, Morarji Desai, Kishorilal Mashruwala, Rajendra Prasad, Ramanlal Desai, Kripalani and Swami Anand.

A weekly magazine, *Jyotiputri* was also published. Initially, the predominant themes of the articles published in it were regarding the status of women in ancient and medieval times and the fight against obsolete customs. Between September 1934 and February 1935, women's political activities were discussed and the importance of physical culture for women was emphasised. In the March 1935 issue, images of women in literature were dealt with. Between April and July 1935, vocations for women and reforms in marriage customs attracted the journal's attention. In July 1938, women's position in law was discussed.

Between 1934 and 1942, Gandhi Week was celebrated every year when cultural programmes concerned with women's awakening were organized; posters mocking obsolete traditions, sexual inequality, child marriage and dowry were sold; plays dealing with social issues were staged and physical exercise camps for women were organized.

Adult education classes were conducted in different localities, including classes for Muslim women. Mridula was very keen that women should be economically independent as she considered this necessary if women were to get an equal status. Jyoti Sangh's Udyog Vibhag taught weaving, soap making, hair oil manufacture, furniture polishing, block printing, sewing, knitting, preparing *papads* and pickles (an idea suggested by Morarji Desai), spices, etc. A small shop was run on the premises which sold articles made in the institution. The Sanchalan Vibhag (where at first the three chief workers were Hemlata Hegishte, Perin Mistry and Charumati Yoddha) undertook rescue work—rescuing women who were victims of child marriage, bigamous marriages, incompatible marriages and prostitution. Mridula held that Hindu Law, as it existed then, gave no rights to women and so she urged the workers of Jyoti Sangh to marry under the Civil Marriage Act and they readily responded. Jyoti Sangh's activities were not limited to Ahmedabad, for it established a wide reputation. Individuals and institutions throughout Gujarat sought its guidance and help. Women trained here were sent to different towns to deal with family quarrels, harassment and physical violence towards women and to give protection to women who had been ostracized by their families, caste or society. There were cases of young women who were heavily insured by their husbands or in-laws and then murdered. Jyoti Sangh raised its voice against this practice. Young widows whose families wanted them to shave their heads often appealed to Jyoti Sangh for help. Mridula or her co-workers met members of these families, wrote letters and after considerable effort usually succeeded in persuading them against it.

She once heard that a *goonda* had locked up a girl and was tor-
turing her. She immediately sent one of her co-workers to go and
rescue her. The co-worker was naturally scared and went with great
trepidation but as soon as the man heard Mridula's name, he said,
'Go and tell your *Pathan* that this girl came here of her own free
will. But if you want, take her away'.[7]

On another occasion, a girl of fifteen wrote that she was being
beaten by her in-laws and so she had run away to her parents. She
had heard that her husband was going to remarry which was against
the Sharda Act and so she wanted help and advice. Or again, there
was an appeal from a young man stating that he had a sister who
was fourteen and had just appeared for her Gujarati vernacular final
examination; she wanted to study further, but their father wanted her
to get married. From Dohad came an appeal to the Jyoti Sangh to
file a case for violation of the Sharda Act, as girls under the
prescribed age were being married off. Girls whose husbands or in-
laws were ill-treating them also came for protection.

Mridula wanted to discourage caste feelings among the women
who came to Jyoti Sangh and so employed a *Harijan* woman,
Manchiben as a peon. She was the daughter of a mill labourer and
was totally illiterate. Many of the high caste women in the Committee
objected to it. Whenever she came they covered their faces and went
home and had a bath lest they had become polluted. But Mridula
stood firm; she gradually persuaded her colleagues and was able to
win them over. For over forty years, Manchiben loyally served Jyoti
Sangh and took part in the Quit India movement. Her daughter
studied upto matriculation and did the teachers' training course.
Mridula also made special efforts to run classes for Muslim and
Bohra women as she was deeply concerned about their backward
condition and was keen on improving their status.[8]

The Jyoti Sangh provided medical treatment to poor women with
the help of hospitals and private doctors. It gave legal aid, made
efforts to provide employment for both educated and uneducated
women and encouraged women to take an active part in municipal,
provincial and Congress elections.

The Jyoti Sangh developed into a unique women's institution and
was followed as a model in other parts of India. It opened the way
for middle class women to obtain further education and do social
service. For women of poor homes, it introduced income generating
schemes so that they could contribute to the family income, and obtain

a respectable position in the family. More important than either of these, it was a refuge for women who were victims of social oppression and injustice.

Child widows like Gangaben Patodia, Gangaben Jhaveri, Jayaben Thakore and others came there and Jyoti Sangh opened a new life for them. It provided a forum through which they could develop their talents and personalities, enrich their lives and take part in public activities. Women from Jyoti Sangh went to work as volunteers in the Haripura Congress. Many of them were active in the Quit India Movement and in Gandhian Constructive Work programmes.

From 1934 to 1939, Mridula paid a great deal of attention to Jyoti Sangh but after that she spent more time outside Ahmedabad and so could not deal with its day to day problems. Nevertheless, she kept in constant touch and helped Jyoti Sangh financially and in other ways. Whenever she was in Ahmedabad, she snatched some time to discuss problems with her colleagues. She and her parents were always giving donations and loans and saving Jyoti Sangh from its perennial financial crises. Her mother Sarladevi was from the beginning on the board of trustees, which included women like Vijayalakshmi Pandit, Lady Tanumati Girjaprasad, Anasuyaben Ramniklal Parikh, Taraben Premchand, Zarina Currimbhoy and Khurshedben Navroji. Mridula had the capability of training ordinary, simple women and she had gathered around her a group of extremely dedicated and loyal workers such as Pushpaben Mehta, Hemlata Hegishte, Charumati Yoddha, Perin Mistry, Udayprabha Mehta, Vidyaben Mehta and others.

Jyoti Sangh was not a political organization but the Congress flag fluttered on its building. When the Governor of Bombay and Lady Lumley visited Ahmedabad in November 1939, the latter wanted to visit Jyoti Sangh and Vikas Griha. The Governor and his wife would not visit any institution which flew the flag of a political party but Mridula refused to take the flag off for them. As a result, they did not visit these institutions.

Jawaharlal visited Jyoti Sangh in September 1939 and was highly impressed by its work. He said that the principal credit for establishing such an institution should go to Mridula whose abilities he had known since she was a child and he was sure that under her able guidance, it would be able to achieve many things.[9]

Jyoti Sangh had soon established a reputation for itself and Mridula received letters from admirers even in distant places like

Nairobi, telling her that Gujarat would be eternally grateful to her for establishing Jyoti Sangh and for her selfless work for the women of Gujarat.[10]

As women in distress started coming to Jyoti Sangh from all over Gujarat for help and protection, something had to be done to provide them with shelter. It was for this purpose that Vikas Griha was established on 1st May 1937 by the joint efforts of Pushpaben Mehta and Mridula. It was a haven of refuge for child widows, battered wives, old women who had no place to go. Vikas Griha had a hostel for girls and a Rescue Home for distressed women and orphans. Jyoti Sangh and Vikash Griha became the two premier institutions in Gujarat working for women.

Mridula took a great deal of personal interest in each individual case in Vikas Griha till 1940. Its main task was to rescue women who had been the victims of rape, abduction, kidnapping, wife beating and other forms of oppression—to give them refuge and help them to regain their self-respect and rightful place in society.

On one occasion, Mridula informed the Police Superintendent that a man living in the Ellis Bridge area of Ahmedabad had locked up his wife and was beating her. The woman had contacted Vikas Griha and informed them that she would commit suicide unless they rescued her. Mridula requested the police to immediately rescue her and bring her to Vikas Griha. Once Bhogilal Lala, the Gujarat Congress leader, found a Gujarati girl who had been turned out of her home, at Poona railway station platform and sent her to Vikas Griha.

In another case a man who had died, had made a will in which he had denied his wife her due share of his property. She appealed for help to Mridula, who immediately took up her case, not only because it affected the lady concerned but because for her, it was a larger question of women's property rights. She wrote for advice to a prominent social worker and lawyer, Motilal Veen: 'This is not just a question of anyone's individual right. It is a question of women's rights and we should not tolerate such acts of injustice'.[11]

Appeals came to Vikas Griha from men who wanted their wives educated and made capable of earning, from girls who were being tortured by parents or in-laws, and often were contemplating suicide; from girls being forcibly married off though they were under-age; from widows or deserted wives or unwed mothers with small babies who had nowhere to go. Mridula never refused help to anyone. Vikas Griha's doors were open to all of them.

To all these women, Mridula's advice always was that they should not suffer such humiliation; they should learn to be independent and self-reliant and then men would respect them.

In the Political Process

In 1936 Jawaharlal Nehru, as President of the Congress, included in his new Working Committee three Congress Socialists, Acharya Narendra Deva, Jayaprakash Narayan and Achyut Patwardhan. He did not, however, include any woman. Sarojini Naidu who had been previously a member of the Working Committee was dropped. Mridula was deeply disappointed that, of all persons, Jawaharlal should have done this and she wrote: You have not been able to select even one woman member for your Cabinet. None of us had imagined that such a situation will arise under your Presidentship. Not only was I surprised at such an omission but I was also pained'.[12] She was not upset at the exclusion of Sarojini Naidu, who, she felt, 'had done nothing actively in the Congress to bring awakening amongst women,'[13] but said that some one else such as Kamaladevi Chattopadhyay should have been taken in. The constitution of the Congress and the resolution on Fundamental Rights adopted at the Karachi Congress in 1931 conferred a status of equality on women. Even though this principle had been accepted in theory, Mridula felt that the Congress had remained indifferent towards the awakening of women. The Congress had made use of women in furthering the cause of the nation but had shown no enthusiasm as regards the creation of their orderly organization and discipline. Further, 'the Congress should remain alive to the problems of women because their condition is worse than that of any class or community. To build up a nation we have been making all efforts; why not make such efforts in the direction of cultivating a noble outlook among men so far as the problems of women are concerned?...I have always hoped that we women who are struggling for our development would get blessings from at least Bapu and yourself'.[14]

She made women members of the Gujarat Congress write to Jawaharlal, protesting against the exclusion of women from the Working Committee and the Parliamentary Committee as 'a direct retrogression from the Karachi Congress creed of equality of opportunity for men and women.... This new and unexpected treatment has pained

us, and shaken our faith in those at the head of Congress affairs...'.[15]
Nehru was not offended by her letter.

I am glad you have written to me as you have given full vent to your enthusiasm for the cause of women. I really appreciate that enthusiasm. I would
like you to realize that my own views and enthusiasm for advancing Indian
women are no wit less than yours. Indeed in many respects they are perhaps
more advanced. You should know that I am not the architect of the new
Working Committee.[16]

Nehru tried to exonerate himself by saying that he had no say in the
selection of the Working Committee and, in fact, he himself was not
happy about it 'I occupied a most unenviable position and circumstances were too much for me'.[17] He implied that Vallabhbhai
and Gandhiji were responsible for the constitution of the Working
Committee and for the exclusion of women from it.[18] When Mridula
wrote to Gandhiji expressing her unhappiness about the exclusion of
a woman member, he denied having anything to do with it and said
that the Working Committee had been constituted by Jawaharlal.[19]
Some misunderstanding was thus created between Gandhiji and
Jawaharlal on this issue and the latter called Mridula to Bombay to
clarify the matter. Jawaharlal, then had to issue a statement that he
took full responsibility for the non-inclusion of women in the Congress Working Committee.[20]

In August 1937, the Congress formed a National Planning Committee with Nehru as its chairman, which appointed various sub-com-
mittees, including one on 'Women's Role in Planned Economy' which
was to consider 'women's socio-economic and legal status, their right
to hold property, carry on trade, profession or occupation and remove
all obstacles or handicaps in the way of realising an equal status and
opportunity for women'. Rani Laxmibai Rajwade was the chairperson
of this sub-committee and Mridula its secretary. The other members
were Sarojini Naidu, Vidyagauri Nilkanth, Vijayalakshmi Pandit,
Radhabai Subbarayan, Amrit Kaur, Perin Captain, Begum Shah
Nawaz Khan, Lilavati Roy, Sarala Devi Sarabhai, Begum Hamid Ali,
Mrs. P.K. Sen and Zarina Currumbhoy. In her usual style, Mridula
set about her new task with enormous zeal and energy, collecting
data and statistics on the position of women in India. A detailed questionnaire was prepared which consisted of seventy questions regarding women's social, economic and legal status, marriage, maternity
and succession, conditions of industrial employment, social customs
acting as hindrances to women, types and methods of appropriate

education and other such problems. Five hundred copies of it were printed and sent to different parts of the country. Provincial convenors were appointed to collect information about women in their own region. The questionnaire was translated into Hindi, Gujarati and other languages.

One of the issues in which there was some difference of opinion was regarding a uniform civil code. While several members wanted a uniform civil code, some Muslim members favoured an optional civil code. Begum Hamid Ali, a member of the sub-committee, felt that no recommendation should be made which would interfere with Muslim Personal Law and wrote to Nehru protesting that some of the reforms proposed were too drastic. Mridula thereupon wrote to Prof. K.T. Shah, Secretary of the National Planning Committee and Jawaharlal Nehru, the president. Nehru advised her to try to come 'to as large a measure of agreement as possible with Begum Hamid Ali in regard to these Muslim matters. Where she feels that your recommendations are against any Muslim injunction, the matter can be dealt with in this way: A few sentences can be added to the effect that some Muslim members of the committee have pointed out that this particular recommendation might infringe the *Shariat* or the customary law of the Muslims'.[21] Nehru felt that the Committee should not impose any recommendation which Muslim women generally considered to be undesirable, or contrary to their religious injunction. There was also some difference of opinion between Hansa Mehta and Mridula regarding the question of women's education. The former felt that women's education should stress on the role of a woman as a mother, whereas the latter felt that women should receive the same education as men and their equality as citizens should be emphasized. The final report of the sub-committee was presented by Mridula on 31st August 1940. The Committee agreed that women's position must be essentially on a footing of absolute equality with that of men in all civic, social and economic matters. It not only reiterated the declaration of Fundamental Rights adopted at the Karachi Congress but went further. Some of its recommendations were quite bold and revolutionary. It recommended abortion for population control and also endorsed the principle of state responsibility for children born out of wedlock. It suggested that marriage must be preceded by a medical examination of both the parties. It also proposed amendment of the civil marriage law to enable marriages between persons intending marriage without any declaration of their religion. Monogamy was to be the law of the land and the right of divorce to be recognized. Co-education was to be promoted and identical standards of morality for men and women were envisaged. Mridula put in an enormous effort in writing this report. The chairperson, Rani Rajwade, acknowledging this, wrote,

'To work as the Secretary of the organization of such vital importance and of so wide a scope was no easy task, more so because for the first time in our national history an attempt was being made to look at things from the women's point of view in its multifarious aspects. You set about the work most systematically. With scientific analysis and precision, you mapped out the entire field of enquiry. For you, no detail was too small to be left out and in fact too insignificant to be missed. With untiring zeal and enthusiasm you attended to the various aspects of the Sub-Committee's work and managed to do everything with characteristic thoroughness even at the risk of great personal inconvenience.'[22]

Pandit Nehru also praised the splendid work that she had put into this report. Mridula said that the work had been a labour of love for her. Following the exclusion of women from the Congress Working Committee as well as from the Congress Parliamentary Board, Mridula decided that women in the Congress must organize themselves. Vijayalakshmi Pandit, Aruna Asaf Ali, Sucheta Kripalani had all expressed the need to establish a women's department within the Congress which would help to create among Congress women an awareness of themselves as a group and a consciousness of their mutual interest and disabilities. Mridula wanted the Congress to take up women's issues seriously which was not easy for even though Congressmen were keen on women joining the struggle for political freedom, very few of them were concerned with questions of women's equality. In fact, most of them were conservative and traditional in their attitude towards women. But she firmly held that it was the responsibility of the Congress not only to fight for political freedom but also for women's rights. She wrote to Gandhiji and Vallabhbhai that the Congress should devote more serious attention to questions pertaining to women. Congressmen should not consider women's issues of no consequence and outside their purview. With this in view, she wrote to the Secretary of the GPCC about forming a Women's Department within the Congress. When the GPCC met on 6th May 1936, it discussed her proposal and decided to appoint a committee under her to draw up a scheme.

The newly formed Women's Committee met at the end of May 1936 and passed a resolution emphasizing the need for a separate organization for women, conducted by women. The Committee framed an outline of the constitution for the proposed body. Before the GPCC approved it, the resolution was circulated for opinion to individual women, women's institutions and those who took interest in such problems. The aim of the organization was to spread the ideology of the Congress among women so that they should take part in the freedom struggle as well as in the constructive programme of the Congress.

Mridula wrote to women workers in different districts of Gujarat as she wanted it to be a grass-root organization. She requested the district Congress committees to gather information on women workers in their areas who could assist them in forming branches of the new organization. Questionnaires were prepared and these together with copies of the scheme were widely circulated among Gujarat Congress members. She was extremely busy from the summer of 1936 until about 1940 in building the Women's Wing within the Gujarat Congress. She wrote letters, issued circulars, and kept in personal touch both with the district-level workers as well as with GPCC leaders like Morarji Desai.

Having started the Gujarat Women's Wing, she next wanted the Congress to have an All India Women's Organization within the Central Congress Office. To the general secretary of the Congress, Acharya Kripalani, she wrote:

The Karachi Congress has explicitly passed Resolutions regarding the fundamental rights of women. Women suffer from various disadvantages and to remove these and to bring women into public life the Congress must provide a space for them.... Women members of the Gujarat Congress would like to introduce a resolution to this effect at the forthcoming Session of the Congress.[23]

She suggested that a conference of Congress women workers should be held at Ramgarh where the next session of the Congress was to be held and wanted the Reception Committee to give them the necessary facilities.[24] She approached Vijayalakshmi Pandit, asking for her and Jawaharlal's support 'as without it, it will be impossible to do anything'.[25] Both Vijayalakshmi Pandit and Mridula issued an appeal to Congress women workers to approach Jawaharlal Nehru and Sarojini Naidu and suggested that Sucheta Kripalani should be in charge of the Women's Department at the centre. The scheme was sent to Dr. Rajendra Prasad, who was then president of the Congress. Mridula suggested that though the Congress had successfully attracted women, it had failed to hold and sustain their interest.[26] Aruna Asaf Ali expressed similar sentiments. In a letter to Rajendra Prasad, Mridula suggested that women had felt handicapped without a separate organization within the Congress which, in turn, had prevented them from effectively contributing to the nationalist struggle.[27] Mridula visualized this Women's Department as being similar to the other departments in the AICC office and to be governed by the same rules and regulations. Rajendra Prasad wrote to her that a new Working Committee of the Congress was soon to be formed and till then no decision on the Women's Wing could be taken.[28] Maulana Azad became the new president of the Congress and she once again approached him, asking him why her proposal was being delayed.[29] On 15th April 1940, when the new Congress Working Committee met at Wardha, it was decided to open a Women's Department in the central office and Sucheta Kripalani was put in its charge.

The women's Department, despite Mridula's enthusiasm and hard work, never really took off. This was partly because of the lack of support it got from Congressmen and partly because soon after its foundation, the Congress launched, first, the Individual Civil Disobedience Movement and later the Quit India Movement. Sucheta Kripalani wrote to Mridula, 'The work of my department is going too slowly for my taste. I have written to all the provinces but the response is not at all good.'[30] Acharya Kripalani asked Mridula to take charge of the Women's Department for a short period but she thought that it would serve no useful purpose to take up the work for just four months. She considered it better that Sucheta Kripalani continue in the post and she was willing to help her.[31] Mridula was willing to look after the Women's Department of the AICC if the office was located in Bombay as she could not leave her work in Gujarat for long. Kripalani wanted the Women's Department to function, like every other department, from the AICC head office at Allahabad. In 1946, after the release of the Congress leaders, the headquarters of the AICC was shifted from Allahabad to Delhi, but the Women's Department was not re-established.

In the Legal Framework

Unlike many others, men and women, within and outside the Congress, Mridula realized that merely by taking part in the Non-cooperation or Civil Disobedience Movement, women had not gained their individual freedom. Emancipation of the country was not necessarily going to emancipate women. In a letter to Acharya Kripalani, she wrote:

If India were to gain freedom today, there is no reason to believe that we will not have to launch a battle for allowing women to live as free human beings. The history of Europe shows that women who participated in revolutions shoulder to shoulder with men, were relegated back to their homes and to their inferior status once the revolution was over. What reason have we to believe that this will not happen in India? Congressmen elected to assemblies are not there only to take up political questions. Is it not also their duty to fight against social injustice and for equality?[32]

Mridula wanted the Congress to take up several legal issues where she believed changes were necessary. One of these was the Hindu Marriage Act. Just as in civil marriage, a notice had to be given before the marriage could take place, so also a notice period seemed to her necessary for a Hindu marriage. Registration of marriages performed according to Hindu rites was also suggested. Such steps, she felt, would lessen child marriage, secret marriages and marriages between young brides and old bridegrooms. If a girl was married under the Sharda Act, her husband was regarded as her first guardian and so even if he ill-treated her, she could not seek assistance from her parents. Rescue of women locked up by their husbands or in-laws or by neighbours was regarded as illegal and police would not rescue them unless the women themselves filed an appeal which was obviously very difficult. Mridula drew attention to these and other lacunae in the legal system which were detrimental to women.

In her fight for women, she often met with resistance from Congressmen, and their conservatism and patriarchal attitude pained and infuriated her. While women workers in the Congress were expected to have high moral standards, the same was not expected of men. She protested against such double standards of morality—one for men and another for women. When *Sandesh,* a Gujarati newspaper, published a report condemning a widow for abandoning her illegitimate child and called it 'an act of sin', she wrote a letter of protest that not a word had been said about the father of the child. Was he not equally guilty? Why should women always be blamed?

* * * *

In the Bombay Legislative Assembly, two social reform legislations were introduced in 1939, the Hindu Divorce Bill by Bhogilal Lala and the Bombay Prevention of Hindu Bigamous Marriages by Lilavati Munshi, wife of K.M. Munshi who was Home Minister of Bombay. Though the Congress Ministry was in office, these were private members' bills and were circulated, as was the practice, for three months to elicit public opinion. Mridula worked vigorously during this period to mobilize opinion in favour of the Bills by writing letters in which she asked her colleagues 'to concentrate for a month or two on a vigorous propaganda... and (organize) demonstrations in favour of the Bills, as it will strengthen the hands of the Congress Government and they will be able to see it through as soon as possible'. She felt, that if passed, the Bills would confer certain rights on Hindu women which were denied to them under the existing laws but she regretted that the Congress Party.had not issued a whip to its members to vote in their favour. 'I think it is high time that the Congress Party in the Assembly takes up this matter and adopts a policy to further measures which will put women on an equal footing with men'.[34] She felt that the Congress could take it up if it so wished, but unfortunately it fought shy of the Bills which had anything to do with the women's movement.[35]

Through Welfare Organizations

In 1941, after the failure to enthuse the Congress about taking up women's issues, Mridula decided to join the All India Women's Conference, which had been established in 1927 in Poona on the initiative of Margaret Cousins.[36] At first, she wanted to join the Ahmedabad branch of the All India Women's Council, but its committee suspected her of wanting to capture the organization and hence she was not welcome. She regarded both the All India Women's Council and All India Women's Conference (AIWC) as organizations run by the upper middle-class women who lived comfortably and wanted to ease their conscience or gain public recognition by doing some social work. She did not have much faith in the way these organizations worked or their objectives. Nevertheless, she joined the AIWC in the hope that she would be able to broaden its base and undertake activities for poor urban and rural women. She drew up a new constitution for the AIWC and sent copies of it to Vijayalakshmi Pandit, Rajkumari Amrit Kaur, Rameshwari Nehru, Kamaladevi Chattopadhyaya, and others asking for their support.

A camp was organized by Mridula at Abhrama in Surat for training women social workers. Some difference of opinion cropped up between her and Kamaladevi (who was the convenor of this scheme) regarding the running of this camp. The latter felt that Mridula was treating it as her own personal scheme and not as that of an AIWC camp. Mridula had her own method of working and would not always take directives from others. Nevertheless, she had put a great deal of effort in organizing this camp.

Mridula, as Organizing Secretary of the AIWC, also had some differences with the Honorary Secretary, Kulsum Sayani and had complained that she was given no freedom to work. She never liked to be in a position where she required permission to do something or where her movement was constrained. She soon found that she could not work in the AIWC and resigned as Organizing Secretary in 1948.

* * * *

Kasturba Gandhi passed away as a prisoner of the British Government in the Aga Khan Palace Jail, Poona on 22nd February 1944. A plan to raise a fund in her name was thought of by Shri Narandas Gandhi of Rajkot. He was supported by Dev Das Gandhi, Shanti Kumar Morarji, Swami Anand, Vaikunthbhai Mehta, Thakkar Bapa and a few others. An appeal for raising the fund was signed by Pandit Madan Mohan Malaviya, S. Radhakrishnan, Purshottamdas Thakurdas, G.D.Birla, J.R.D.Tata, Ambalal Sarabhai, Kasturbhai Lalbhai, Mridula Sarabhai, Sarojini Naidu and many others. In all, about 100 persons of all shades of public opinion put their signatures to this appeal which was issued on 8th March, 1944. A target of collecting seventy-five lakhs of rupees was envisaged by 2nd October, 1944, the next birthday of Mahatma Gandhi. In fact, a crore of rupees was collected before that date.

On his release from jail on 6th May 1944, Gandhiji was requested and prevailed upon to accept the chairmanship of the Kasturba Gandhi National Memorial Trust (KGNMT). He felt that as Kasturba had been a simple woman, devoted to village life, the object of the fund should be the welfare and education of rural women and children. The Board of Trustees, at a meeting held on 1st July 1944 at Poona, resolved to coopt ten more trustees, and authorized the chairman, Mahatma Gandhi, to suggest the names. Among the persons he nominated as trustees was Mridula. For better administration and smooth running, the Board of Trustees resolved to have an executive committee with Mahatma Gandhi as chairman, Sir Purshottamdas Thakurdas as vice-chairman, and A.V. Thakkar, popularly known as Thakkar Bapa, as secretary. Mridula was a member of this committee also. At the meeting of the Board of Trustees held in November 1944, Mridula was appointed organizing secretary by Gandhiji.

A central office of the organizing secretary of the KGNMT was set up on the premises of Scindia House at Ballard Estate in Bombay. Since Gandhiji desired that as far as possible women should be associated with the working of the Trust, she recruited mostly women in the Central Office and selected persons who had either been associated with the freedom struggle or with the constructive programme of Gandhiji. She also made it a point to appoint women of different communities. Amy Moos, a Parsi lady who had provided shelter to underground leaders and workers during the 'Quit India' Movement, as well as a nationalist Muslim lady, Irshad Fatima, were selected. Among the others who worked in the Central Office were Kamala Patel, Leela Jog, Madhavi Narayanan, Kali Prasad and Dhote. The recruitment of all the members of the staff had the approval of Thakkar Bapa, the general secretary of KGMNT who had done pioneering work among the tribals and aborigines of Gujarat, was president of the Harijan Sewak Sangh and for whom Gandhiji had great regard and respect.

Differences developed between Mridula and Thakkar Bapa regarding their method of working, their attitude towards women, the location of the office and generally about the aims of the organization. According to Mridula, while Thakkar Bapa emphasized relief work, KGNMT's aim was to give women self-expression, and to create a 'new' woman. This ideal he did not share. He was old-fashioned, traditional and could not appreciate strong, independent, self-willed women. She wrote to Gandhiji that as long as Bapa was the general secretary, women were not going to be allowed to be in command or take policy decisions. While she had great regard for him, she considered him a misfit for this job. The first clash occurred regarding the appointment of the office secretary. Thakkar Bapa wanted to appoint Shyam Lal, who was serving as secretary of the Harijan Sevak Sangh at Delhi, as he felt that only a man could handle this work. Mridula did not agree and wanted Pupul Jayakar for that post. Thakkar Bapa, however, was not prepared to appoint Pupul Jayakar. He wrote to the vice-chairman of the Trust, Purshottamdas Thakurdas that Gandhiji wanted Shyam Lal appointed as full-time secretary in the central office of the Trust. The trustees thereupon decided to appoint Shyam Lal. In order not to offend Mridula's feelings, Gandhiji suggested that Pupul Jayakar could be requested to work in an honorary capacity, and she agreed to do so.

One of the main activities of the Trust was to establish hospitals, charitable dispensaries, child-welfare centres, leper colonies and homes for women and children in the rural areas of India. To start the work, an Advisory Medical Board was formed consisting of eminent doctors including Dr Jivraj Mehta and Dr Susheela Nayar.

Special training was required for rural workers. Mridula, therefore, started an All India Training Camp in April 1945 in Borivili, Bombay, under the auspices of KGNMT. She was overall in charge of the camp and took great personal interest in planning the programme in every detail and in its execution. She wanted the women to wear the Punjabi dress, *kurta* and *salwar* as she felt that it would make their movements less restricted. She also wanted them to take off their glass bangles which many wore as a sign of being married, during physical exercise. Such rules were often resented.

Mridula had some differences of opinion with a few organizers of the camp on whether prayer should form a part of the daily programme and should be made compulsory. She was opposed to this idea and held that attendance at prayer meetings should be voluntary. The matter was referred to Gandhiji, who came and delivered a speech at the Borivili camp. 'Prayer should be a spontaneous welling up of the heart', he said, 'one should not pray if one felt that prayer was a burden'.[37]

Mridula, in her capacity as organizing secretary of the Trust, toured the villages of Rajasthan, Madhya Pradesh, Bihar, Karnataka, Gujarat and other provinces. She wanted to be in close touch with the provincial representatives of the Trust. Pupul Jayakar recalls how she accompanied Mridula on many of her tours and how this was her first exposure to Indian villages. Mridula, according to her, was an excellent guide and teacher.[38] She wanted the branches to organize training of *gram sevikas* and start various programmes for the welfare and education of rural women and children.

There was no specific demarcation of the executive powers of the general secretary and the organizing secretary of the Trust and this led to frequent misunderstandings. Both the secretaries began to feel that the other was overstepping the jurisdiction of his/her executive power. As days went by, the differences began to widen. Thakkar Bapa's style of working was quite different from that of Mridula's. She was always in a hurry to do things and wanted to bring about a radical change in the outlook of the village women. While Mridula was young, energetic and forceful, Thakkar Bapa was older, more cautious. careful and methodical. He was very meticulous about the financial aspects of a scheme and wanted accounts to be kept carefully. Gandhiji's business acumen and efficiency were almost legendary and he appreciated Thakkar Bapa's way of working. Thakkar Bapa did not approve of Mridula's methods, speed, haste and impatience. When she sent telegrams, he felt letters would have sufficed. Thakkar Bapa made no secret of his differences with her and discussed them with his colleagues. The matter was ultimately brought to the attention of Gandhiji and each of them told him that he/she could not work with the other. Thakkar Bapa offered to resign but Gandhiji asked him to continue[39] and asked Mridula to proceed on leave for a short time. Mridula wrote long letters to Gandhiji and the latter wrote to Thakkar Bapa 'I have received a pile of papers from Mridulaben'.[40]

The office of the general secretary was shifted from Bombay to Wardha, but Mridula continued to function for some time from Bombay.[41] Differences between Thakkar Bapa and Mridula however continued and so she submitted a letter of resignation to Gandhiji in July 1945, requesting him to relieve her from the post of organizing secretary. Gandhiji said that their assessment of Thakkar Bapa differed, he knew him longer than she did but that did not necessarily mean that he was right and she was wrong. It would be presumptuous on his part to think so. He held that it was wise for her to resign since she and Thakkar Bapa could not pull along as they were temperamentally so different and he felt that Thakkar Bapa was necessary for the organization. So he allowed Mridula to resign as organizing secretary but wanted her to continue as a trustee.[42]

Thus ended Mridula's eight months official connection with the KGNMT. When she resigned, her aunt Anasuyaben was most upset and expressed her feelings in a letter to Gandhiji who replied, 'I appreciate your down-heartedness and devotion to Mridula. No one can help being devoted to her. Such is her work, sacrifice and bravery. But you are mistaken if you feel that she will be lost to the cause. The result should be quite the contrary.[43]

Mridula continued to take interest in the work of the KGNMT. In October 1949, she invited Madam Aung San, the widow of the first prime minister of independent Burma, to Sevagram to attend the tenth KGNMT meeting. She wrote to Begum Abdullah, the Maharani of Patiala, the Begum of Rampur and several other ladies inviting them to visit Sevagram and see for themselves the institutions being run by the Trust. In December 1948, when an editorial appeared in the *National Herald* which was critical of the Trust's work, she wrote a letter of protest to the editor, Chelapathi Rau.[44] She pointed out to him that it was the only institution doing work exclusively on an all India level in villages through women workers.

In January 1960, the Board of Trustees set up an Assessment Committee to suggest future plans for the Trust. Mridula discussed the problem with Vinoba Bhave and put forward her proposal for women's role in developing *lok shakti* (people's power) and *lok neeti* (people's code) in the *sarvodaya* pattern of society. She wanted to rouse *stree shakti* (women's power) and believed the KGNMT was an appropriate forum for this. She prepared a note regarding the re-organization of the Trust activities and placed it before the Assessment Committee. The Committee found the suggestions extremely valuable and invited her for personal discussions. She felt that there was an urgent need to provide a cadre of full time women social workers on the pattern of the Servants of India Society (founded by G.K. Devdhar and others in Poona in the 1880s) and wanted the Trust to concentrate on the training of such workers. Her continuing interest in the work of the Trust despite the fact that she was removed as its organizing secretary, reflects her lack of pettiness and her genuine concern for the work which the Trust had undertaken for improving the condition of rural women.

CHAPTER SIX

COMMUNAL RIOTS

Ahmedabad 1941

Ahmedabad has always been prone to communal riots. So on 17th April 1941, when riots broke out, it was not unexpected. Rumours had been circulating since March of the possibility of clashes between Muslims and Sikhs. Tension was mounting as *lathis*, knives and daggers were being brought into the city. The Muslim League had been propagating communal hatred since the breakdown of the Congress–League negotiations for coalition ministries in Bombay and U.P. in 1937. Mosques were being used for this virulent propaganda. Volunteers were trained to use weapons and the *Khaksars*[1] were active. There had been a few minor incidents, including an assault on the *chaukidar* (guard) of Vikasgriha and the murder of a Sikh, over a money lending transaction in Raipur a Hindu dominated locality. Two nationalist Muslims had been attacked on the 16th of April. The police had been aware of this but no action had been taken to prevent the out-break of violence. Administrative apathy—deliberate or due to inefficiency—was one of the causes for the out-break of this communal riot. Mridula claimed that she had proof that the riots had been instigated by the Pathan branch police officers under the orders of their British superiors. This was to punish the citizens of Ahmedabad for not contributing to the war fund. She also claimed that men had been brought from Dacca and other places to create trouble.

Mridula was in Bombay when she read in the Gujarati newspaper *Janmabhoomi* that riots had broken out in Ahmedabad on the previous day. She immediately rushed back to find the city completely paralyzed. The streets were deserted, all transport was off the road, even telephones and the ambulance services were not functioning. On the night of Friday, the 18th, some Muslim *goondas* had set fire to shops at the crossing of Char Rasta in Khadia and in Manek Chowk market in the middle of the city. Houses and shops had been looted, burnt and destroyed so that Manek Chowk, Gandhi Road, Khadia Char Rasta, Pancha Kuva all presented the appearance of bombed areas. The loss to Hindu life and property was enormous but they could do nothing immediately except take shelter behind barricades. The office of *Sandesh*, a Gujarati newspaper had been set on fire. Then, the Hindus retaliated. In this first phase of the riots, which lasted ten days, nearly eighty persons were killed and around four hundred were injured. Property worth Rs 15 crores was destroyed and about one lakh persons left Ahmedabad. Streams of refugees could be seen, with their luggage on their heads, on the roads leading out of the city. The textile mills were closed for one week and as a result workers lost their wages. People were afraid to move out of their houses without police protection. No Hindu could enter Kalupur and Jamalpur, where Muslims were in a majority and where Hindu shops and houses had been looted and burnt. Temples had been burnt in Shahpur, Manek Chowk and Jamalpur. Similarly, no Muslim could pass through Khadia and Raipur which were Hindu areas.

In her characteristic manner, without wasting time, Mridula went to Raipur which was a Hindu stronghold and tried to reason with the leaders. She pleaded with them to refrain from looting and burning shops or attacking mosques. But no one was prepared to listen to her. They asked her where she had been the previous day when Muslims started the trouble—'First go and talk to the Muslims, then come to us.'[2] The few Muslims residing in this area had left. On the day the riots broke out, as Pushpaben Mehta, Indumati Seth and Puratan Buch were taking the last batch of Jyoti Sangh workers home in Mridula's car, a Muslim mob armed with *lathis* (sticks), knives and hockey sticks surrounded them and started throwing stones. Pushpaben was injured but as a result of the driver's courage and presence of mind, further injury was avoided. On another occasion, Indumatiben displayed great courage when she appealed to a mob advancing towards Jama Masjid not to indulge in violence. When she went into the middle of the crowd, people feared that she might be killed but her appeal found response and the mob withdrew.

Mridula met Congress leaders including Bhogilal Lala, and the secretary of the GPCC. She wanted their approval to form a committee to guide the citizens in restoring peace and to organize Congress volunteers who would do relief work and go around explaining to the people the importance of maintaining communal harmony. Congress leaders felt that all this was going to be useless, no one would respond to their appeals, but seeing Mridula's enthusiasm and persistence, they allowed her to go ahead. She began operating from Congress House with the help of a few Congressmen like Khandubhai Desai, Chimanbhai Patel, Prabhudas Patwari and Puratan Buch. Khandubhai Desai and the members of *Majoor Mahajan* (The Congress Trade Union) did excellent work during the riots. They kept the ambulance service going, the telephone in Congress House was kept functioning, an enquiry office in the Civil Hospital was opened, arrangements were made for the disposal of dead bodies, protection was given to those who wanted it and help to people moving from one area to another.

Mridula wrote to Sardar Patel[3] asking for instructions and a long letter to Gandhiji[4] in which she traced the entire history of the riot. She also expressed her unhappiness over the way in which the Gujarat Congress leadership had behaved. She said that she was not trying to be unfair to anybody nor accuse anyone but was just pouring out her troubles and if she could not turn to him at such moments of crisis where was she to go?

She felt that the Congress had not done enough during the riots. The Muslim League on the other hand, had encouraged communal violence. In Astodia the local League president, Mr Bukhari and secretary, Mr Momin were present when every car which passed was systematically attacked. The majority of inspectors in the police force were Muslims, and so they did not take any action against the Muslim mobs and kept the Collector in the dark. The Hindus had retaliated in an equally despicable manner. She appealed to Gandhiji for help and asked him to send Mahadev Desai to Ahmedabad to give them guidance.

Gandhiji was deeply concerned about the inactivity of the Congress amidst all this communal hatred and violence, and commended the courage shown by three women—Mridula, Indumati Chimanlal and Pushpaben Mehta, who at the risk of their own lives tried to restore peace. He expressed his deep anguish at the outbreak of the communal violence and emphasized how the Congress was the only organization which from the very beginning was trying to build a united India. But if Congressmen adopted a communal attitude, all its efforts would be wasted. He regretted the marginal role played by Congressmen during the riots in maintaining communal harmony. It was Mridula who had fully briefed Gandhiji about the Ahmedabad riots over four days both through correspondence and discussions with him as well as Gulzarilal Nanda.

Mridula was depressed by the innocent lives that had been lost, houses and shops that had been burnt but most of all by the apathy of the people—particularly Congressmen, who seemed to be bothered only about the votes in the next elections and power politics. The Congress was a divided party and its organization seemed to have collapsed during the riots. She was critical of the Hindu bias shown by the Congress leadership of Gujarat.

In order to restore communal harmony, the Shanti Sevak Sangh (Peace Workers' Union) was formed with Mahadev Desai as president, Narhari Parikh as vice-president and Bhogilal Lala and Gulzarilal Nanda as secretaries. Among the members were Mridula and her aunt Indumati Chimalal. To Morarjibhai, Mridula suggested that just as there were the All India Spinners' Association and the All India Village Industries' Association, so also there should be an All India association for maintaining communal peace and harmony. To Jawaharlal she wrote: 'I am glad to inform you that due to the little training I received from Bapu and you, I was able to keep myself well composed during this hour of trial.'[5]

In May, when there was a fresh outbreak of communal violence, many of the Congress leaders were in jail because of the individual *Satyagraha*. During this period, Mridula was appointed for four months as one of the secretaries of the City Congress Committee. She moved about in the riot affected areas trying to remove fear and restore peace. She submitted regular reports to the committee formed under Mahadev Desai's chairmanship. Since riots were often the result of rumours, the Shanti Sangh started an information department and the City Congress Committee gave it support.

In 1946, *Rath-Yatra*, a Hindu festival and *Id* a Muslim festival were to be observed on the same day. As was expected, communal feelings were aroused and riots broke out. Prominent Congress Muslims were attacked and, as in 1941, *lathis*, daggers and knives were smuggled into the city. The *Rath-Yatra* procession started, as it did every year, from the Jagannath temple, and taking the usual route, it reached Kalupur circle near Raj Mahal Hotel where it was stoned. In the ensuing communal riots, Mridula was once again active in trying to restore peace. In the riots, two fearless young workers, Vasantrao Hegishte and Rajab Ali, were killed. A small committee was formed consisting of Khandubhai Desai, Arjun Lala, B.K. Dalal,

Jayanti Thakore, Kantilal Ghia and Mridula to formulate a plan for lessening the communal tension in the city.

When for the first time during a communal riot an attack was made on two old women, a statement strongly condemning this was issued signed by Vidyagauri Nilkanth, Mridula and several Hindu and Muslim women of the city.

Meerut 1946

Twenty days before the Congress session was to be held in Meerut in 1946, there were communal riots in the city. People believed that the Muslim League was bent on not allowing the Congress Session to be held peacefully. At that time, Jawahrlal Nehru, the president of the Congress, was abroad and Mridula who was one of the general secretaries of the Congress, was in Delhi. She rushed to Meerut instead of Allahabad, where she was originally scheduled to go, and met the district collector accompanied by the U.P. Congress Committee chairman, Seth Damodar Swaroop and secretary, Sarjoo Prasad.

On 8th November 1946, a special train carrying Hindu pilgrims from the annual fair at Garhmukteshwar was attacked and looted and the passengers manhandled at Dasna railway station, seven miles from Ghaziabad and about twenty miles from New Delhi. Several persons were injured. The engine driver left the train at the mercy of the mob who ransacked it, looted railway property and set fire to the station building. The mob was, however, finally beaten off by the passengers. Pandit Govind Ballabh Pant and Rafi Ahmed Kidwai arrived on the scene. A unit of troops came from Meerut and took charge of the wounded, including a number of women suffering from severe burns, stab wounds and effects of criminal assault.

On hearing the news of this attack, the remaining Hindu pilgrims, on their way back from Garhmukteshwar, who had to pass through Muslim villages, gathered together in a large convoy. The majority of them were Jats from Rohtak and Hissar districts (Haryana). A strong rumour gained currency that these Jats on their return journey, would burn all the Muslim villages that were on their way. On the other hand, there were possibilities that trouble could also be created by the Muslims in the roadside villages. Meerut city was in panic and signs of trouble were apparent. Transport services were failing because drivers were afraid to drive. When Mridula heard this, she contacted the deputy commissioner of Meerut and apprised him of the situation. She personally took the responsibility of maintaining communal harmony and peace in the area through which the convoy was to pass.

The next morning, Seth Damodar Swaroop, Sarjoo Prasad and Mridula went around by car to study the situation and see for themselves what arrangements had been made by the government. They could not find any police on the way. As they passed through Hindu villages, they got down and advised the people to maintain peace in their areas. When they reached Shahjahanpur, a predominantly Pathan village, with nearly half of the population Muslim, they found the people terror-struck. They had weapons of all kinds for self-defence and were praying to God for His mercy. Mridula and her colleagues talked to the Pathans and assured them that they would not be attacked by the Hindu pilgrims, but advised them not to provoke the Hindus in any way. Then they went to the only telephone available six miles away to request the deputy commissioner to immediately despatch an adequate police force.

As they were returning to Shahjahanpur, they saw a group of jubilant Jats carrying *lathis*, spears, axes and swords as if they were preparing for an attack. Damodar Swaroop and Mridula got down from their car on which the Congress flag was flying. The Jats were also carrying Congress flags and raising slogans like—Mahatma Gandhi *ki jai*; Jawaharlal Nehru *ki jai*; Congress *Zindabad*; *Musalmano ko khatam karo* (Finish off the Muslims) *Khoon ka badla khoon'*. (Blood for blood).

Mridula went into the midst of this unruly crowd. As she was too short to be seen, she climbed a tree-trunk so that the people could see her properly, and started pacifying them. She advised Sarjoo Prasad to go back to Meerut to inform the authorities and request them to send a heavy contingent of police immediately. He reached the office of the Reception Committee set up for the Congress session and said: 'I cannot say whether Mridula is alive or dead.' In the meantime, she was explaining to the crowds that while shouting 'Mahatma Gandhi *ki jai ho*', they were really stabbing him in the back. For full seven hours she and Damodar Swaroop struggled with the mob, preventing it from moving forward, preaching and trying to persuade the people that the path of violence was suicidal. They also assured them that the Pathans would not attack them and that a police force was arriving shortly to protect them. One of the youths in the mob shouted that she was a Muslim and Jinnah's sister. Damodar Swaroop shouted back that her name was Mridula Sarabhai and she was the general secretary of the Congress. The youth, however, insisted that she was the sister of Jinnah and had come to save the Muslims. Damodar Swaroop had a hard time persuading him that she was not Fatima Jinnah, but Mirdula Sarabhai. She on her part told the mob that Fatima Jinnah had never worked for Hindu-Muslim unity and would never have the courage to face a Hindu crowd. This had a magical effect on the crowd and a section of them raised the slogan—'Mridulaben *Zindabad*'.

Shah Nawaz Khan, who had accompanied Mridula on her visit to some of the villages, wrote to Jinnah about this incident and how rank communalists and hooligans were trying to incite both Hindus and Muslims. He mentioned how 'Mridula Sarabhai, when trying to bring home the futility of senseless destruction of life and property was insulted by Hindu pilgrims incited by RSS and other communalists.'[6] He added that she deserved to be congratulated for her courage and determination, in spite of the fact that the police had tried to discourage her from approaching these mobs.[7]

Dusk had fallen but no contingent of police arrived and people were getting impatient and restless. All of a sudden a few vehicles were seen approaching. In one of them was Pandit Govind Ballabh Pant, Chief Minister of U.P. and Chaudhary Charan Singh. They were naturally very concerned about Mridula's safety and had brought with them an adequate police force. As a result of this the Hindu pilgrims' convoy passed through the Muslim villages peacefully.

Damodar Swaroop, Sarjoo Prasad and the others would not have gone to Shahjahanpur, had they not been coaxed and pressurized by Mridula. It was indeed fortunate that she happened to be in Meerut when news of the outbreak in Shahjahanpur reached her. She continued to stay in Meerut preparing for the forthcoming session of the Congress. In order to prevent communal riots, she reorganized the Congress Seva Dal under the command and guidance of Major General Shah Nawaz Khan and Lt. Col. Bhonsle of the Indian National Army. She issued a note regarding the spread of communalism within the Congress and of the need to distinguish between pseudo-Congressites and true Congressmen.[8]

Noakhali 1947

On 15th August 1947, when the transfer of power took place in Delhi, Gandhiji was in Noakhali in East Bèngal (Pakistan) where there had been ghastly communal riots, to restore confidence among the uprooted Hindus and to heal the deep wounds sustained by them. Asking Hindus to shed fear, he said.

Fearlessness radiates through the atmosphere... It has the qualities of a magnet. One brave person gathers around him other brave men. A whole group thus shakes off its fear and begins to take part in activities which might formerly have cost their lives.[9]

Mridula accompanied Jawaharlal to Noakhali to have a glimpse of the riot-torn villages and see for herself the miracle Gandhiji was bringing about. During the two days she spent there, Mridula tried to study the problems of the Hindus and the technique adopted by the Mahatma in rehabilitating them in their abandoned villages. Together with Jawaharlal, Acharya Kripalani and Shankar Rao Deo, she was present at Gandhiji's prayer meeting on 28th December 1946, where he said that the Congress was not a Hindu organization and he had come to prove by his actions that he was a sincere friend and well-wisher of Muslims. Restoration of the feeling of amity and brotherhood was his sole concern. After the prayer meeting, at Haimchar, Mridula had a meeting with Gandhiji late one night. He asked her to accompany him to Bihar where mass killing of Muslims had taken place as retaliation for Noakhali. Mridula joined Gandhiji's *Shanti Sena* movement in 1946 during this visit to Noakhali.

Bihar 1947

Hardly had the communal riots in Noakhali subsided, when trouble broke out in Bihar. Muslims attacked Hindus in Calcutta and Noakhali, the Hindus in Bihar retaliated.

The Calcutta 'Direct Action', of August 1946, set up a double shock wave, one travelling in the direction of Noakhali, the other in that of Bihar.... More than a million Biharis earned their livelihood in Calcutta as shopkeepers, rikshaw pullers, door-keepers, etc.[10]

A large number of them had been killed in the riots following the 'Direct Action' day on 16th August. Till 1911–12, Bihar formed a part of Bengal presidency, and as a result, there was a large Bengali population in Bihar whose relatives and friends had been killed in

Calcutta and East Bengal. The survivors who returned to Bihar as refugees carried harrowing tales. An atmosphere of panic was created and a rumour spread that the Muslim League would repeat the massacre of Hindus as it had done in Calcutta and Noakhali. Mysterious leaflets started circulating, many of them obviously under faked names, others anonymous, containing instructions to kill Hindus.[11]

Towards the end of September 1946, an incident occurred at Benibad in Muzaffarpur district, which sparked off the riots. A local Muslim was reported to have abducted a Hindu girl from Calcutta and the people demanded her return. The Muslim promised to do so in two or three days. On the appointed day, a crowd went to his place but found that the man had disappeared and there was no trace of the girl. The enraged crowd killed a few Muslims and burnt their homes.

Soon after, communal riots spread to the districts of Chapra, Patna and Gaya. About three and a half lakh Muslims were reported to have fled from Bihar to different places, men and women were brutally killed and even children and infants were not spared. 'Noakhali Day' was observed and over four hundred villages were completely destroyed. Gandhiji rushed to Bihar from Noakhali. His task in Noakhali had become difficult in view of the massacre of Muslims in Bihar. He wanted to prove by his action that all victims of this insane communal frenzy, Hindus or Muslims were equally dear to him. He had originally intended to go to Bihar only for seven to ten days. Five months had elapsed since the Bihar carnage but the situation was still tense and at the request of the Governor, Dr Syed Mahmud, he decided to stay on.

Gandhiji asked Mridula to accompany him and she kept a diary of those days entitled 'Gandhi in Bihar, 1946–7' which was found accidently thirty three years later in her mother's papers and published as *Pratham Pratyaghat*. The others who were associated with Gandhiji in the relief and rehabilitation work in Bihar were Dev Prakash (brother of Pyare Lal), Manu Gandhi, Hamid Hunnar, his Urdu translator, and Professor Nirmal Kumar Bose. Tension had gripped both the Hindus and Muslims and few persons were prepared to go to Bihar and work in an atmosphere steeped in hatred and suspicion. Khan Abdul Ghaffar Khan and Shah Nawaz Khan reached there to help in this task of restoring communal peace. Mridula worked as Gandhiji's secretary for a brief period from 5th to 29th of March, and was with him when he was passing through a period of great mental torture and anguish. He was staying with Dr Syed Mahmud, in his house on the banks of the Ganges where from morning to night, thousands of visitors poured in. In Noakhali the complainants had been Hindus; in Bihar they were Muslims. Women walked with their children from distant villages to have a *darshan* of the Mahatma whom they regarded as a saint. As soon as the *ramdhun* and *bhajans* were over, many left the prayer meeting or created disturbance or noise during his speech as they did not understand what he was saying. Mridula accompanied Gandhiji as he moved from village to village on foot in scorching heat, speaking directly to the people in the villages and as he put it to 'read in the face of the country-side the mystery of what had happened'.[12] He strove to restore confidence among Muslims, asking Hindus to atone for their sins, wiping tears from the eyes of the homeless and bereaved. His speeches after the prayer meetings preached love and non-violence. He repeatedly said that even if the Muslims of Noakhali and Calcutta had behaved abominably, Hindus of Bihar should not take revenge by killing Muslims. He kept on urging Muslims to return to their homes. Those who wanted to meet Gandhiji usually first met Mridula and she examined their problems and questions. If Gandhiji had already provided the answers, she conveyed them to the visitors and if there were new problems, she took them to him. She led advance parties which preceded Gandhiji's arrival in a village, cleaning up, making all the preparations. In the heat of May, she kept Gandhiji's living room as cool as she possibly could by using blocks of ice and *khas tattis*.[13] Often she had to spend five to ten hours arranging a public meeting for Gandhiji but she said that she benefited enormously from this work. She learnt a great deal about human psychology, why people behave in different ways during communal riots and also Gandhiji's methods of tackling problems in dealing with human beings. It was a great experience for her to observe how he was giving strength and confidence to those who felt weak and helpless, advising them to fear no one but God and remove every trace of fear from their hearts. From Gandhiji, she learnt to be fearless and the importance of faith in oneself and courage of convictions. 'If one has inner strength, he needs no army to protect him.' She now understood the meaning of Tagore's song. 'If no one listens to your call, walk alone.'[14] Mridula was deeply impressed by Gandhiji's way of resolving the communal problem during the Bihar tour and was convinced that restoring Hindu-Muslim unity was a very important area in which she should continue to work even after Bihar. Gandhiji was also appreciative of her work: 'Mridula has been accompanying me... she is working very hard, and she does not discriminate between a Hindu and a Muslim,' he said in one of his after prayer speeches.[15]

Ahmedabad 1969

The riot that began on 18th September 1969, was one of the worst in Ahmedabad's history. It was unprecedented in terms of size and proportion of people involved and the intensity and speed with which it spread. Official figures calculated the casualties as 660 dead and 1400 injured, while rumours put it at 3000 killed. The riot was linked to events of the preceding eight months. In the municipal corporation elections of January 1969, two Jan Sangh leaders were elected from a traditionally left-oriented ward. The Jan Sangh was naturally jubilant and the Muslims very concerned. Prior to the municipal elections, the RSS had held a camp in Ahmedabad which was addressed by Guru Golwalkar. Since then, there had been minor incidents. On 18th September, a trivial incident took place involving some *sadhus* and a group of Muslim boys near Jagannath Temple, outside Jamalpur gate, to the north of which lived a large Muslim population.[16] This small incident was broadcast as disastrous and the rumour was spread that 300 Muslims with acid bulbs had attacked Jagannath temple and wounded several *sadhus* seriously. That night a Muslim washing shop was set on fire and it was alleged that the police deliberately did not help. It was also said that a Hindu mob attacked a mosque, looted it and set it ablaze. Minor events of this kind followed one another.[17] The newspapers gave a highly biased account and instead of cooling the situation, fanned the flames of communal hatred. The following day, hand-bills were circulated carrying an exaggerated account of the Jagannath temple incident and alleging that the Muslim leaders had refused to apologise. This was not true since the Muslim leaders had offered their apologies and also assured their co-operation in the efforts to restore peace. The *sadhus* were angry and refused to accept any apology. The Hindu *Dharma Raksha Samiti* held a meeting where provocative speeches were made. By this time the city was in the grip of strong communal fever. An organized group of around 200 to 300 people fully armed with *lathis*, spears and flaming torches marching through several lanes attacked Muslim houses, shops, mosques and *darghas* in a planned and systematic manner. The mob was full of wrath and the city seemed to be at its mercy. Muslim shops were broken open, goods were looted and shops set on fire. The fire brigade was nowhere on the scene to rescue the shopkeepers. The worst affected were the labour areas. *Chawls* comprising rows of huts were completely destroyed.

Mridula was in Srinagar when she got news of the riot and on 22nd September she arrived in Ahmedabad and immediately met the Chief Minister Hitendra Desai. The Congress had split in 1969 into the Congress (I) under Indira Gandhi and the Congress (O) with Morarji Desai. Hitendra Desai belonged at that time to the latter faction. He welcomed Mridula's offer of help and suggested that she should work in the refugee camps that were being set up. This suggestion was endorsed by the Revenue Minister, Premji Thakur and the Gandhian worker, Ravi Shanker Maharaj. She, therefore, started working at the refugee camp at the police stadium. Initially her various suggestions for running the camp were accepted but soon all the decisions taken by her were reversed and the camp was handed over to the police. She was informed that she could not work in the camp without the permission of the chief minister. She tried desperately to meet Hitendra Desai and Premji Thakur but neither of them was willing to meet her.

In the Indo-Pakistan war of 1965, Balwantrai Mehta, the then chief minister of Gujarat, was killed as his plane was shot down by Pakistan. Many Gujarati Hindus had been bitterly against Pakistan since then. Rumours were afloat that Mridula Sarabhai was an agent of Sheikh Abdullah who was pro-Pakistan. Sheikh Abdullah had celebrated 26th September as a protest day against the Ahmedabad riots and this was attributed to reports sent by her. Strong protests were made against her role in Ahmedabad by several ministers, Congressmen and even, it is said, by the governor of Gujarat, Sriman Narayan and his wife.[18] Mridula, thereupon, returned to Delhi and complained to Morarji Desai and Hitendra Desai about the unfair accusations made against her. She appealed to Jayaprakash Narayan for support.[19]

The Ahmedabad riots and the communal violence was a challenge to Indira Gandhi who launched a frontal attack on the RSS. The Congress (O) in Gujarat suspected Mridula of being an agent of the prime minister who, it was alleged, was trying to encourage the anti-Hitendra Desai faction in the Gujarat Congress.

By early November, Jayaprakash Narayan, Khan Abdul Ghaffar Khan and Narayan Desai wanted Mridula to return to Ahmedabad and work with the *Shanti Sena* in its efforts to restore communal peace. Morarji Desai also told her that he did not doubt her bona fides nor did he hold her responsible for what was happening in Kashmir. He saw no objection to her working and continuing her efforts towards restoring communal harmony in Ahmedabad. She was naturally much relieved and reassured by this trust reposed on her and started working among the riot-affected women and children, mostly Muslims, who were staying in refugee camps or in improvised shelters on the streets. Narayan Desai helped her start a women's section, the *Stree Seva Vibhag*, in the *Shanti Sena* to rehabilitate women and children. About three hundred women had lost their husbands or earning members of their families in these riots and many more such women were coming to Ahmedabad from surrounding areas for shelter and to file their claims. Nearly 1,200 children were homeless and there were many orphans.

Mridula was shocked to find that many people in Ahmedabad, including some government officials, were opposed to the idea of a transit camp for these homeless people who were apparently told that the government would be hostile to them if they helped her and the *Shanti Sena's* efforts. The latter found it very difficulty to obtain accommodation to start a camp and efforts to start the camp with the co-operation of relief organizations having failed, it decided to launch this programme as a women's movement. They were aided in their efforts by Dr Rahmatullah, Ghulam Rasool Qureshi and workers of Jamiat-ul-ulema-Hind. They were able to persuade Sunni Muslim Waqf authorities to give them the use of their *Zenana Yatim Khana*. Funds were assured by Jamiat-ul-ulema-Hind, the Central Relief Committee and the *Shanti Sena* and so the transit camp finally started on 17th January 1970. The opponents started spreading rumours and frightened women away from joining the camp. Mridula and her co-workers launched a counter campaign and started an economic rehabilitation project for these women in the camp.

The starting of the camp three months after the riot, despite all the opposition was a victory for Mridula. She was a person never to give up easily, and believed that obstacles had their own reward. These hurdles led her to contact women of all sections of Muslims and non-Muslims and enabled her to renew many old acquaintances and come in touch with new forces, which had influence on their society and community. She found this a rewarding experience and hoped to develop these contacts in future.

The camp was being run by Hemlata Hegishte and Gulshan Shroff but Mridula continued to take interest in it from Delhi. She met Maulana Abdul Rais, and Maulana Yusuf, president and secretary of Jamat-i-Islami respectively and explained to them the problems she faced with their workers in Ahmedabad. She wanted them to actively cooperate with the *Shanti Sena* and particularly with its women's wing which was running the transit camp. The Muslims of Ahmedabad recognized her efforts.

We cannot forget the work which you have done in Ahmedabad especially for the widows who were not cared for and the timely help from you had considerably raised the high hopes for the future of the widows. We shall not forget the efforts you have made in Ahmedabad and I hope that your efforts will be successful.[20]

In January 1970, efforts were under way to set up an organization at the instance of Abdul Ghaffar Khan on the lines of the *Khudai Khidmatgars* (Servants of God) to promote communal harmony and revive Gandhian ideals. Mridula was actively involved in the efforts to start the organization which was named *Insani Biradari* (Human Brotherhood). Two of its objectives were, to eschew violence, and encourage the spirit of tolerance and mutual respect among all the people of India in regard to each other's religion, aspects of culture and way of life. A convention was held on 16–17th August 1970 in Delhi, and Jayaprakash Narayan was elected its president; Sheikh Abdullah, senior vice-president; Badruddin Tyabji, Kunwar Mahendra Singh Bedi and Khalil Ahmed, vice-presidents; Mrs Raksha Saran, treasurer; Shah Nawaz Khan, general secretary and Mridula Sarabhai and Radhakrishnan, secretaries. The *Biradari* functioned from a room in the Gandhi Peace Foundation. The *Sarva Seva Sangh* of Vinoba Bhave advanced Rs 20,000 for the organization. However, the movement soon ran into difficulties. Shah Nawaz Khan wished to resign as general secretary on being appointed a minister at the Centre in May 1971.[21] One of the vice-presidents, Khalil Ahmed, disapproved of the manner in which decisions at the meetings were being taken.[22]

Mridula met Abdul Gaffar Khan and related to him the difficulties confronting *Insani Biradari*. He had received 'a purse' during his visit to India in the Gandhi Centenary year. There was a general feeling that he should donate this money to the *Insani Biradari*. But he wished to disassociate himself completely from the organization on the ground that during communal riots, *Insani Biradari* was not able to give any help to the victims and its office bearers had not even issued a statement condemning the riots. Jayaprakash had, according to him, issued a statement, without proper enquiry, blaming Muslims for the riots. In the Ahmedabad riots, not a single person had been brought to book. 'It is said that the arm of the law is powerful to reach the real culprits. What are your laws good for then, I ask? The laws are for the preservation of peace and justice, not a cover for violence and wrong.'[23] He further elaborated how ineffective the *Insani Biradari* had been in realizing its aims. 'I don't think it is beneficial to Hindus and Muslims. Then where is the question of giving funds? I don't think this can defend Muslims. So where is the need for such a movement.'[24] However, he suggested that Jayaprakash and Sheikh Abhullah should work in close cooperation; the former should work among Hindus and the latter among Muslims, try to draw them away from communalist feelings, create mutual good-will and thus pave the way for the interaction of the two communities.[25]

It soon became clear that many members did not want Mridula and Sheikh Abdullah in the organization as they felt that their presence was hampering its work. Mohinder Singh Bedi wrote to her that wherever they went to collect funds, people hesitated to contribute because the two of them were involved. Mridula, therefore, asked for leave from the *Insani Biradari* in September 1971.[26] The same week she wrote to Shah Nawaz Khan (who had continued as general secretary after joining the cabinet, despite his earlier letter of resignation) and expressed her strong protest against the lavish dinner hosted by him, as general secretary, to a delegation from Bangladesh. She argued that office-bearers should not spend on entertainment, when they were so short of funds. She reminded him that the Movement was inspired by Badshah Khan to implement Gandhian ideals of communal harmony.[27] In her letter to Mohinder Singh Bedi, she pointed out that the work of the organization had not suffered in any way due to her presence or that of Sheikh Saheb.[28] She continued to be associated with the *Insani Biradari* till March 1972 when her leave expired, following which she resigned. The organization was in deep financial crisis and ceased to be effective any longer.

Through all these years, Mridula responded equally to Hindu and Muslim suffering in communal riots. From Gandhi and Nehru she had learnt not to discriminate and also that it was the duty of the majority community to remove the fears and protect the minorities.

CHAPTER SEVEN

THE GREAT MIGRATION

The Partition of India in 1947 and the resultant migrations and massacres represented a human tragedy of enormous proportions. It 'enforced movements of people on a scale absolutely unparalleled in the history of the world.'[1] 'There must be many examples in the bloody history of mankind where the extent of violence has been as great or even greater but it is probably true that there has never been such a big exchange of population,' wrote Horace Alexander.[2] It is estimated that about five and a half million people travelled *each way* across the new India-Pakistan border in Punjab. As Sir Francis Moodie, Governor of West Punjab wrote to Jinnah 'the refugee problem was assuming gigantic proportion'.[3] As an historical event, partition had ramifications that reach far beyond 1947, yet historical records make little mention of the dislocation of people's lives, the strategies they used to cope with loss, trauma, pain and violence. Many women at that time took up relief work among refugees through their own or their family's involvement, through contact with some important person or as a result of personal loss or tragedy.

After getting permission from Gandhiji to witness the Independence Day celebrations in Delhi on the 15th August 1947, Mridula left Patna where she had been working with him for restoring communal amity and peace in Bihar. At the Red Fort where the national flag was being hoisted, she learnt from a journalist friend that communal riots had broken out in Punjab. She hurriedly contacted Jawaharlal Nehru and expressed her desire to proceed to Punjab immediately. Nehru was hesitant to allow her to go to Punjab without obtaining approval of Mahatma Gandhi. But when news began to pour in that the law and order situation in both the Punjabs had gone out of control, he himself suggested that she should proceed to Amritsar at the earliest. 'I felt honoured that I had been entrusted with this difficult and responsible job. Jawaharlal informed Gandhiji on the phone that I was going to Amritsar.'[4] She had planned to go in her own car but was told that the Grand Trunk Road from Delhi to Amritsar was blocked by violent mobs. Nehru advised her to travel by a plane up to Lahore and contact the Inspector General of Police, Khan Qurban Ali Khan, who had served in U.P. and knew many Congressmen. Lahore was quite new to her. She had visited it only once in 1929 when the famous annual session of the Congress was held there and Nehru took the pledge of complete independence on the banks of river Ravi. She tried to contact Qurban Ali Khan but having failed to do so, started making enquiries about how to go from Lahore to Amritsar. While doing so, she saw a convoy of Hindus and Sikhs on its way to Amritsar and asked whether any of them could accommodate her. Someone recognized her and requested her to get into his vehicle. All persons in the convoy were terror-stricken and in a state of panic as they were not even sure they would reach Amritsar alive. They were thus happy to have her with them. Her work for Muslims in Bihar was known by them and they hoped that her presence would deter Muslim mobs from attacking them. Mridula did her utmost to raise the morale of the frightened women and children and throughout the journey tried to keep them in high spirit. The moment the convoy crossed the border at Attari in East Punjab, and entered India, a full throated cry of '*Hindustan Zindabad*' broke out.

On reaching Amritsar, she found that anarchy prevailed all round. There was a complete breakdown of communications as well as of law and order. Everywhere there was an atmosphere of panic and fear. Half the population of both East and West Punjab seemed to be on the roads looking for refuge. Only a few days had passed since Partition and both the governments were desperately trying to cope with the millions of refugees crossing the border. The administrative paralysis caused by the reshuffling of cadres on a communal basis and the infiltration of communalism into the police and military had, by the end of August 1947, created a situation in which it seemed impossible for Hindus to stay in West Punjab and for Muslims to stay in East Punjab. The governments of East and West Punjab had to face a terrible crisis in the very hour of their birth, even before they had settled down to work or had proper offices functioning.

On her arrival in Amritsar, Mridula got in touch with the deputy commissioner, Nakul Sen, who invited her to stay in his bungalow. On 17th August, a high level meeting took place in Lahore between the representatives of India and Pakistan which included the prime ministers of both the countries as well as the governors and premiers of West and East Punjab. After a prolonged discussion, it was agreed that the evacuation of refugees should be undertaken by both the Punjab governments. The governments of India and Pakistan undertook the responsibility of maintaining train services and communications for their evacuation.

After their meeting in Lahore, Nehru and Liaqat Ali Khan, prime ministers of India and Pakistan respectively, came to Amritsar, met the officials and went round the riot affected areas of the city. Mridula, together with the local leaders and some government officials, accompanied them. The visit by the prime ministers created an impression among the public that peace would be restored but soon after their departure, the city was once again plunged into turmoil.

It was difficult for Mridula to decide how to begin her work and from where to find workers. The situation was terribly complex and difficult. She had no idea how to proceed except what she had gleaned from her experience in Bihar.

She started her work from the Amritsar Hotel on the Mall Road, which had been the best hotel in the city. The proprietor of the hotel, a wealthy Muslim, had fled to Lahore on the day communal riots broke out in Amritsar. The building was requisitioned by the government for relief work and six rooms were allotted to Mridula for her offices. for the evacuation of refugees and the recovery of abducted women.

Amritsar was full of Hindu and Sikh refugees, who had suffered greatly in Pakistan—who had been victims of the orgy of riots, loot and arson, and many of whose family members had been killed, raped or abducted and who had lost their houses and all their property. There was understandably a great deal of bitterness against Muslims. In such an atmosphere to work for Muslims was neither easy nor popular.

Thousands of Muslim refugees were stranded at the Amritsar railway station on their way to Pakistan. Mridula selected this as one of the venues for starting her peace mission and she posted a group of workers there to prevent RSS volunteers from harassing them. It was not uncommon for members of one community to mock, jeer and poke fun at the cowed, hungry and miserable evacuees of the other community. She also posted a second batch of volunteers inside the walled city of Amritsar to protect Muslims who were isolated and to keep her informed of day to day developments. On the basis of their reports, she asked the deputy commissioner and the senior superintendent of police to deploy more police force for their protection. As a result, the leaders of the Muslim associations in Amritsar began to look to her for support and consulted her frequently. She continually put pressure on the police and civil administration to act promptly to safeguard the Muslim population in Amritsar. But many local Congressmen, RSS as well as Shiromani and Akali Dal leaders disapproved of what they regarded as her pro-Muslim attitude. She felt that the Punjab unit of the CPI was the only political party which stood for the protection and safety of Muslims and got the support of their leaders like Sardar Sohan Singh and Sardar Teja Singh. When one of them, Makkhan Singh Tarsikka was arrested under the Public Safety Act by the East Punjab Government, her reaction was sharp. She immediately wrote to Gopalaswamy Iyengar, Minister without Portfolio in the central government, saying that Tarsikka was not engaged in any subversive activity but was her co-worker who had completely devoted himself to relief work. When Sardar Gehal Singh and four other communists who had played an active role in protecting Muslims and escorting them to Pakistan were found missing, probably killed by militant communalists, she was terribly upset and once again asked Gopalaswamy Iyengar to intervene.

She deputed batches of social workers at places where Hindu and Sikh refugees were pouring in and was in touch with Jawaharlal on telephone and informed him of the plight of these refugees. Here, of course, volunteers of the Rashtriya Swayam Sevak Sangh and Shiromani Gurdwara Prabandhak Committee and many others were also doing excellent work.

Mridula's main function was to try and protect the Muslims in Amritsar and East Punjab and arrange for their safe evacuation, while at the same time to help rehabilitate the Hindu and Sikh refugees who were streaming in from West Punjab. She kept the Prime Minister informed of the developments in Punjab and Nehru visited Amritsar for a second time on 24th August, 1947 to discuss the situation with leaders and workers of various political parties. He addressed small road-side gatherings, appealing to the people that retaliation in any shape or form was no remedy and that it was the duty of Sikhs and Hindus in East Punjab to protect the minorities as that was how the interests of the minorities in West Punjab could best be protected. He was not in favour of the wholesale migration of population.

Conditions worsened in Pakistan by early September and leaders of all political parties began to feel that it would be folly for Hindus and Sikhs to stay on. Bhim Sen Sachar, who was the leader of the Congress Party in the West Punjab State Assembly, along with his colleagues took refuge in Amritsar. On arriving, he contacted Mridula and narrated his personal experience and expressed his view that it had become necessary to evacuate all Hindus and Sikhs from West Punjab as quickly as possible. In September 1947, it was decided at the third meeting of the Emergency Committee of the representatives of India and Pakistan that the situation in Punjab had deteriorated so much that mass evacuation of Muslims from East Punjab and Hindus and Sikhs from West Punjab had become imperative. An agreement was reached to facilitate the migration of population from both the Punjabs. The central and provincial governments were to assist in this task. Mridula was unhappy at this decision, for she was keen that Muslims should continue to live in East Punjab in amity with Hindus and Sikhs.

To evacuate the Hindu and Sikh refugees from West Punjab, one hundred trucks were placed at the disposal of the deputy high commissioner of India in Lahore. To help refugees to recover their relatives as well as their property, the East Punjab government appointed a chief liaison officer at Lahore, and a transport controller at Amritsar. Similar arrangements were made for the evacuation of Muslims from East Punjab. But these arrangements offered no protection to the refugees, because there was no machinery to prevent attacks on the convoys from hostile mobs. It was, therefore, decided to entrust the work of evacuation of refugees to the military. Two Military Evacuees Organizations (MEO) were established, one each at Lahore and Amritsar, to evacuate in the shortest possible time the maximum number of Muslims from East Punjab and Hindus and Sikhs from West Punjab. The refugees had to be moved under military protection to transit camps and from there to refugee camps. To cope with this gigantic and complex problem, a new Ministry of Refugee Relief and Rehabilitation was created by the Government of India and K.C. Neogy was appointed its minister. A Cabinet Emergency Committee was appointed to deal with the highly volatile situation.

Panic-stricken Hindu and Sikh refugees started coming to the Amritsar Hotel pleading for the early evacuation of their relatives who were stranded in West Punjab and NWFP. They were afraid that otherwise they would be butchered, or their girls abducted and forcibly converted to Islam. Many of them insisted on meeting Mridula, as they had been told that she was the personal representative of Jawaharlal Nehru. She had a team of social workers among whom were Dr Prakash Kaur, Dr Upkar Singh, Makkhan Singh Tarsikka, Dr Bhatia, Professor Khushpaul Singh and a few others who helped her. Towards the end of September 1947, a group of Congress Socialist leaders who were ardent supporters of Jawaharlal Nehru arrived in Amritsar from Rawalpindi and Lahore along with Bhim Sen Sachar. They joined Mridula in this herculean task. Thus, a very strong group of political and social workers was formed in Amritsar.

Mridula knew almost all the political leaders of Pakistan. including the rehabilitation minister of West Punjab, Mian Iftikhar Uddin who had been in the Indian National Congress till 1946. He introduced her to the Central Minister of Pakistan, Raja Ghaznafar Ali Khan who helped her in the evacuation of Hindu and Sikh refugees from West Punjab. Mridula's contribution in helping the Muslims at Garhmukteshwar (Meerut district) and Bihar was fresh in the memory of Pakistan leaders. As such, they always cooperated with her in her mission. One of the main tasks at the Amritsar Hotel office was to contact the district liaison officers by telephone and get information regarding the situation in each district. On the basis of this, the office of MEO was requested to deploy a convoy of military vehicles to a particular district which needed urgent attention. Lala Avtar Narain Gujral and Dr Lehna Singh Sethi, liaison officers in Gujarat and Sargodha (West Punjab) contacted her by telephone after 9 p.m. every night and apprised her of the position in their districts. She used to contact Major General Chimney, head of the MEO in East Punjab and request him to make speedy arrangements for the evacuation of refugees. By directly contacting the highest military officers of the MEO, she could get things done speedily.

Evacuation on a massive scale began from West and East Punjab towards the end of August and early September 1947, through special train services. With this began the attacks on refugee trains. Passengers were forced into compartments like sheep and goats and because of the heat, they often found it hard to breathe. In the ladies' compartments, women would try to calm down and comfort their children. Looking out of the window, they could see corpses piled up on top of each other.[5] Attacks on refugee trains were carried out with a military precision which had perhaps something to do with the role played by ex-servicemen, Hindus and Muslims, of the Indian army. While revenge and bloodthirst were the main motives, violence was also used to ensure ethnic cleansing. On 21st August, a train carrying Hindu and Sikh refugees was attacked between Wazirabad and Godhra. On 28th August, Sikh refugees from Sialkot were attacked near Narowal railway station. An attack on a train carrying Hindus and Sikhs at Pind Dadan Khan on 21st September, 1947 in which nearly 1,500 persons were killed, caused tremendous commotion in Amritsar. In retaliation, Hindus and Sikhs attacked a train, carrying Muslim refugees, on the outskirts of Amritsar. Mridula had warned the officials that such an attack was being contemplated and arrangements should be made to avert it. She was, therefore, very sore that no preemptive action was taken. She was firmly convinced that if the local police and military had come to the assistance of the military escort of the train, the attackers could have been repulsed. Throughout the night between the 22nd and 23rd of September, she, along with a group of workers, attended to the wounded and the injured, amidst the groaning and wailing of women and children. With the help of officials, she made arrangements for sending those who were still alive to Lahore by motor vehicles.

Mridula brought this incident to the notice of Nehru and suggested that the deputy commissioner and the senior superintendent of police be transferred from Amritsar at once for dereliction of duty and this was done.

As refugees from Pakistan were pouring into East Punjab and Delhi in large numbers, the government set up camps for them. There were about 7 lakh Hindus and Sikhs staying in refugee camps at Amritsar, Cheharta, Gurdaspur, Ferozpur, Jullundar, Ludhiana, Rohtak, Hoshiarpur, Hissar, Ambala, Karnal, Panipat, Shahabad, Kurukshetra and Gurgaon. The refugee camp was like a small township. As far as the eye could see there were tents. Hundreds of men, women and children milled around. Refugees arrived all through the day and night in bullock carts and trucks under armed protection. People did not dare set foot outside the gates without police protection. Shelters were provided wherever possible. All schools and colleges in these areas were closed and all available buildings and space requisitioned. Tents were also put up. The Indian National Congress appointed a Central Relief Committee with Sucheta Kripalani as its secretary and also the United Council for Relief and Welfare (UCRW) was set up with Lady Mountbatten as its chairperson. The latter achieved the phenomenal feat of uniting twentyone Christian organizations under the banner of the UCRW which set up many centres in East Punjab including one at Amritsar.

Mridula took a keen interest in organizing training and production programmes at these centres and for this, enlisted the cooperation of many experienced Gandhian constructive workers, the foremost of whom was Bibi Amtus Salam. The secretary of the UCRW, B.N. Banerjee, relied on Mridula for advice and guidance on various matters pertaining to social work. In order to get first hand information about the condition of the refugees, Lady Mountbatten together with Mridula visited the transit camps in Amritsar and Jullundur districts. The uprooted women and children presented special problems, which required urgent attention. In order to solve them, many women's organizations mobilized active workers. In particular, the All India Women's Conference and YWCA contributed substantially. To formulate and enforce a proper policy on the rehabilitation of women refugees, the government appointed a small Advisory Committee of women social workers in the Ministry of Rehabilitation. Rameshwari Nehru, who had been looking after the evacuation of women and children from West Punjab during the worst disturbances, had come to Delhi in November 1947. She was invited by the government to become honorary director of this committee which was responsible for the care, maintenance and rehabilitation of uprooted women and children from Pakistan. Mrs John Matthai and Mrs Hannah Sen worked as honorary secretary and honorary joint-secretary respectively. Among those who helped them were Mrs Shobha Nehru, Mrs Raksha Saran and Dr Sushila Nayyar.

To rehabilitate all the Hindus and Sikhs who had migrated from West Punjab, NWFP, Baluchistan and Sind was a tremendous task. Equally difficult was the rehabilitation of those Muslims who had been driven out from their homes in India but were not willing to migrate to Pakistan. Among such Muslims were the *Meos* of Mewat in the district of Gurgaon who claimed their descent from the Rajputs. When a convoy of *Meos* from Delhi and Alwar proceeding to Pakistan was attacked enroute, Mridula argued fiercely with the authorities that they should be given full protection to stay on in India. In those highly surcharged times, when Hindus and Sikhs in Pakistan were being massacred, her action was considered by many to be 'quixotic and almost unpatriotic'.[7] An eye for an eye and a tooth for a tooth was the popular feeling. But Mridula was a firm believer in the Gandhian idea that minorities must be protected. This was Nehru's policy also. The Muslims of Buria, Alwar, Bharatpur and Chamba did not want to migrate to Pakistan but could not return to their homes for fear of being attacked by Hindus. Mridula deputed a group of Hindu social workers to go to Buria to try and arrange for the return and rehabilitation of these Muslims through the agency of the UCRW. In this work, Mridula and her colleagues faced the hostility of the local populations and it was only because of her zeal and perseverance, helped by the Gandhian constructive worker Satyam Bhai, that these Muslims could stay on.

The problem of resettling the Muslims of Chamba also attracted Mridula's attention. Chamba formed a part of the district of Kangra which is now in Himachal Pradesh. Formerly, it was a part of East Punjab. With Mridula's initiative, a centre was started there by the UCRW to resettle the Muslim *Gujars* who had been driven away from their homes following Partition.

Mridula was also very anxious to start constructive work on Gandhian lines in Jammu and Kashmir State. She arranged a network of activities there under the auspices of the UCRW for the rehabilitation of Hindu refugees who had been uprooted from their homes in the Pak-held territory of J & K state. She set up a women's home in Champiari for the rehabilitation of unattached women, children and the physically handicapped. A social worker of the UCRW was deputed to be in charge of classes for spinning and weaving and the medical dispensary. Rehabilitation centres were also set up at Rajauri, Sunderbani, Mender, Nowshera, Banduwala, Ranbirsinghpur and Jasrota.

Another gruesome attack on Hindu and Sikh refugees was made on 23 September 1947 at Jassarh, a border town in the district of Sialkot. The refugees were being carried in a train which was escorted by the Pakistan military. On reaching the railway terminus, they were ordered to get down and walk on foot across the bridge on the river Ravi for entering into India. The military escort had gone back. While they were trying to cross the bridge, an attack was made on them by a huge mob of Muslims. On hearing this, Mridula instantly drove to Jassarh in Sialkot District (Pakistan) and was received by the deputy commissioner of Sialkot and Pakistani police officers. A large number of girls had been abducted by the mob and over 500 persons had been killed and 430 wounded. She made arrangements for removing the wounded to hospitals and transporting the survivors to Amritsar.

The news of this attack at Jassarh created a sensation in Amritsar. Infuriated Sikhs and Hindus began to pour into the city from the adjoining villages armed with swords, spears, daggers, guns and all kinds of weapons, wanting to take outright revenge. A few thousand Muslim refugees were camped at the cattle fair ground on the outskirts of the city. Mridula contacted the police and military and suggested that more men should be deployed to avert an outbreak of violence. At her suggestion, the entire community of Muslim refugees was cordoned by the police and army. Sardar Narinder Singh, Deputy Commissioner, Chaudhary Ram Singh, Senior Superintendent of Police and Brigadier Chopra personally supervized the operation. Mridula, the SSP and DC explained to the Sikh and Hindu mob that instructions had been received from the Prime Minister and the Defence Minister, Sardar Baldev Singh, that the Muslim refugees had to be protected at any cost. Eventually the mob melted away. Lala Bhim Sen Sachar was heard commenting, 'to-day we saw Bahenji in the role of a lioness'. Though the police and army were responsible for controlling the frenzied crowd, it was Mridula who had taken the initiative in preventing a counter-attack on the Muslim refugees.

The cattle fair ground incident of Amritsar convinced Mridula that had the police and the army been deployed with strict discipline and honesty in both India and Pakistan, communal disturbances could have been averted and the exodus of minorities from East and West Punjab could perhaps have been avoided.

This episode increased her credibility in Pakistan where she was referred to as '*Mureed-e-Allah*'—servant of God. Her courage as well as her non-communal approach were admired and Pakistan officials listened to her, regarded her as a valuable link with India and gave her full cooperation. In India also, Nehru, many Gandhians, and a few others appreciated her work. She was invited by Lady Mountbatten to accompany her on a tour of West Punjab. Together they visited the refugee camps at Lahore, Lyallpur, Muzaffargarh, Dera Ghazi Khan, Sialkot, Rawalpindi and Kasoor, where Muslim immigrants from East Punjab, Patiala and other places had arrived. A report of this tour was prepared by Mridula. Lady Mountbatten was full of praise for this, describing it as 'admirable, very balanced as well as constructive'. She wrote:

I would like to tell you what a great pleasure it was to me to have you with us on our tour and how grateful I was for all your sympathetic understanding and help. Do not forget that we shall be so happy to offer you hospitality here at any time and I shall rely on you to let me know when you come to Delhi so that we can put you up at Government House'.[8]

Lady Mountbatten and Mridula also toured East Punjab, visiting Amritsar, Amabala, Ferozepur, Gurgaon and other places to gain first hand information about the condition of Hindu and Sikh refugees.

In a speech in the Security Council on 15th January 1948, Zafarullah Khan alleged that the Ahmadiya headquarters and his house in Qadian had been destroyed by Hindus and Sikhs. The Ahmadiyas were a heretical Muslim sect whose founder Mirza Ghulam Ahmed of Qadian claimed to be a prophet. Since the Koran held that Mohammed was the last prophet, Mirza and his followers were declared heretic by orthodox Muslims and the Pakistan government. Mridula together with the Minister for Rehabilitation, K.C. Neogy and General Thimayya visited the town of Qadian.[9] It had a predominantly Muslim population, hence Muslims from neighouring areas had taken refuge there. From nearly 17,000, its population increased to 70,000 and this in turn caused panic and tension among the local Hindus and Sikhs. The majority of Ahmadiyas wanted to migrate to Pakistan but a few wanted to stay behind. Neogy, Thimayya and Mridula held consultations with a deputation of Ahmadiays led by Nazeer Ahmed Sahib and also with Hindus and Sikhs of the town. It was decided to provide adequate protection to Muslims who wanted to migrate.[10] Mridula followed this up by a visit to Lahore and discussed the Qadian question with K.L. Punjabi, ICS, Officer-on-Special Duty in the office of the Deputy High Commissioner of India in Lahore. They met Khalifa Basheer Ahmed, Supreme Head of the Ahmadiya sect, who complained that they were being forcibly evicted from Qadian. On her return from Lahore, she learnt that the situation in Qadian had deteriorated considerably as a unit of the RSS was openly preaching violence to oust the Ahmadiyas. She passed on this information to the Deputy Commissioner, General Thimayya and the Governor of East Punjab. As a result, precautionary measures were taken promptly and nothing untoward happened.

In Lahore, she met a number of Muslims who had been with the Congress before 1947 and were unhappy in Pakistan. They wanted to come to India but the East Punjab government was reluctant to help them. She reported this fact to Nehru and wanted the government of East Punjab to help them to cross over but her efforts met with little success. During this time, she was often at odds with the East Punjab government. She accused it of having a communal bias and of not taking adequate and firm measures to suppress the spirit of retaliation among Hindus and Sikhs. According to her the rank and file of government officers as well as the police were partisan and ministers were anxious not to lose votes in the next elections. In a note, she wrote that the Chief Minister, Dr Gopichand Bhargava was weak and what they had in Punjab was not a true Congress but an Akali government. She added that preparations were under way to have *Sikhistan* and private armies were being raised and suggested the dissolution of the Punjab Congress Government and the imposition of section 93.

This note did not find favour either with the Congress Working Committee or the Punjab Government and the latter was anxious to remove her from East Punjab. She was dubbed as a 'Pakistani' agent and a 'traitor' but she could not be removed as long as she had the backing and support of Nehru. In a letter to Mohanlal Saxena who had taken over as Minister of State for Refugee and Rehabilitation, Nehru enclosed a note from Mridula Sarabhai who had written about the Punjab government taking possession and disposing of lands and houses of Muslims who had temporally vacated them without going to Pakistan. He asked Saxena to frame a line of action by consulting people like Vinoba Bhave, Gopalaswamy Iyengar and Mridula Sarabhai.[11] Mridula was reassured when the AICC passed a Resolution reiterating that it did not accept the two nation theory and that the rights of the minorities would continue to be protected.[12]

Hindu and Sikh refugees continued to arrive in truck loads from Pakistan. Many of them went to Amritsar Hotel and requested Mridula to put pressure on the MEO to deploy convoys of lorries to particular pockets where their relatives were stranded. She verified their stories and tried her best to help the genuine cases. It had become quite common for the refugees to raise slogans such as 'Mahatma Gandhi *ki jai'* and 'Jawaharlal Nehru *ki jai'*, '*Bahenji* Mridula *ki jai'*. Once, however, when she found a report to be false and refused a request for immediate evacuation of some refugees, the crowd became very hostile and shouted 'Mridula *Bahenji murdabad'*. They squatted to meet her but she refused as they had shouted slanderous slogans.

A train load of Hindu and Sikh refugees were to be repatriated from Bannu, NWFP to India via Lahore. The train could come either via Gujrat (in Pakistan) or Sargoda. Gujrat was still in turmoil and taking the train that way was hazardous. The deputy high commission of India was told that the train would arrive in Lahore via Sargoda, but it did not. It was deliberately diverted at the last moment and taken to Gujrat where the passengers were brutally looted, raped and massacred. News of the tragedy reached Lahore but representatives of the deputy high commission of India were told not to go to Gujrat and even the high commissioner was not allowed to go. Mridula had returned to Lahore from Delhi on the previous night and the moment she got news of the attack, she contacted Sri Prakasa, the Indian High Commissioner on the telephone. He assured her that he would do the needful and cautioned her not to take any undue risk. She then got in touch with Khan Qurban Ali Khan, Inspector General of Police of West Punjab and told him that she was proceeding to Gujrat. He had great regard for Mridula and told her not to come as they were making arrangements for bringing the injured passengers to Lahore. But this did not satisfy her and she asked him either to provide her with a police officer as an escort or warned him that she would proceed on her own'. Qurban Ali Khan was taken aback on hearing this but knew the futility of arguing with her. He reluctantly provided two police officers to accompany her. They drove throughout the night and reached Gujrat early next morning to find the station strewn with corpses and wounded men women and children. She immediately took charge of the situation, got the dead, wounded and alive separated, arranged milk for the surviving babies, saw that the wounded were removed to Gangaram Hospital in Lahore and the rest were taken to the nearest transit camp for Hindu and Sikh refugees at Gujranwala. The survivors were heard saying that the Goddess herself had come to their rescue in the form of Mridula. She did not leave the station till every dead person was identified and the remains disposed off according to Hindu rites.

When Mridula arrived in Amritsar from Lahore, a group of refugees who belonged to Bannu flocked to Amritsar Hotel wanting to know the names of the abducted and dead persons. She told them that nobody could answer this question since the survivors were in a state of terrible shock and tried her best to pacify them.

Jawaharlal wrote to Rajendra Prasad that he had received infor- mation from Mridula that the situation in Amritsar had deteriorated greatly, specially following news of the Gujrat train incident. Sikh gangs were moving about and threatening reprisals. There was con- siderable apprehension that these Sikh gangs would attack not only trains carrying Muslim refugees or stores to Pakistan but also raid

Pakistani villages. Sir Chandulal Trivedi, Governor of East Punjab wrote several letters to Sardar Swaran Singh, the Home Minister of East Punjab, that armed bands of Sikhs were operating in practically all the districts of Jullunder Division thus making the restoration of law and order difficult.[13] Sir Francis Moodie, Governor of West Punjab wrote to Sir Chandulal Trivedi complaining about a well organized plan to exterminate Muslims and drive them out of East Punjab.[14] Swaran Singh replied that there was no organization behind these lawless activities and Sikh leaders were trying to restore law and order.[15]

Mridula received information that two refugee trains had already left for Amritsar carrying Muslim passengers and that a plan was afoot to take revenge for the Bannu episode by killing the passengers of these trains at Amritsar. She therefore telephoned from Delhi to Kamlaben Patel who was working closely with Mridula in Lahore and asked her to rush to Amritsar, contact the district collector and ask him to make the necessary arrangements for the safety of the passengers on these trains. When Kamlaben conveyed this message to him, he was not very happy. But Mridula wielded so much authority in those days that officers were afraid to ignore her instructions. On her way back from the collector's house, Kamlaben saw a huge crowd trying to remove the rail tracks near Amritsar railway station, but timely intervention averted the disaster. The episode revealed Mridula's ability to act quickly and firmly under critical conditions.

In one refugee camp, Mridula found a group of children utterly bewildered and dazed. When she talked to them, they started crying and said that they wanted to be shifted. She tried to console them and assured them that alternative arrangements would be made. She called one of the social workers and instructed him to make the children comfortable and send her the bills for any extra expenditure involved. The children were nicknamed in the camp as children of Miss Sarabhai. One of them, Surender, had been studying in a school in Jhelum before partition. When Mridula found out that his schooling had been discontinued she made arrangements for his education at a school in Benares and requested Rao Saheb Patwardhan to act as his guardian. Through Rao Saheb, she kept in touch with the academic progress of the boy in the years to come. This was just one of many such incidents.

There were nearly 3,000 Muslims who lived in Faridabad and were engaged in *Mehandi* (henna) cultivation and were forced to leave due to the pressure of events. They had drifted to Delhi or other parts of India and some had even gone to Pakistan. Some of them were now anxious to return. Nehru was in a quandary because the houses and lands of these Muslims had been allotted to Hindu and Sikh refugees from western Punjab. While about 1000 Indian Muslims who had temporarily left Faridabad wanted to come back, it was difficult to dislodge the newly settled Sikhs and Hindus unless adequate arrangements were made for them elsewhere.[16] In the rehabilitation of refugees in Faridabad, Mridula played an important role, particularly in carrying messages to Nehru and getting some crucial decisions quickly made.[17]

Hindu and Sikh refugees from West Pakistan, narrating tales of woe and atrocities, caused an explosion of communal strife in Delhi in early September 1947. A wild rumour gained currency that the Muslims of Delhi, under the direction and guidance of the Muslim League, would overthrow the national government of independent India and capture power in Delhi. This did not seem absurd to all Hindus at that time as almost all the ammunition dealers in Delhi were Muslims, and they were well-trained in the technique of manufacturing country-made bombs and other lethal weapons. The rapid increase in the number of Hindu and Sikh refugees in the city began to unnerve the Muslims who started to leave their homes in areas where they were in a minority and began to congregate in the *mohallas* where they were in a majority. Some of them also began to collect firearms and other weapons for self-defence.

The situation became extremely tense and incidents of sporadic firing and hooliganism started within the walled city and the outer areas of Karolbagh, Sabzimandi, Sadar Bazaar and Kashmiri Gate. It became dangerous for Muslims to visit areas which were populated predominantly by Hindus and, likewise, Hindus did not dare to visit Muslim localities. Though curfew had been imposed in many parts of the city, it was not effective and rioting started in an organized manner, specially in those areas which had a mixed population. Stabbing, murder, arson, loot and abduction of women became a matter of common occurrence. Bloodthirsty ruffians broke into houses and dragged the men out and stabbed them. The civil administration seemed to have completely collapsed and it was difficult to maintain law and order. The capital of India was paralyzed by one of the worst communal riots in its history. Connaught Place became the target of attack and almost all the shops owned by Muslims were looted. On hearing this, Prime Minister Nehru rushed there, and getting down from his jeep, with a small stick in his hand challenged the mob which was looting the shops. He declared that violence would be suppressed sternly and troops were moved into the city. Inside the walled city, in the glow of fire from burning buildings, gunshots and the shrieking of women and children, along with deafening cries of *'Allah O Akbar'* and *'Har Har Mahadev'*, rent the air all night long. The Muslims were so terror-struck and demoralized that they did not know whether to leave their homes or stay on. They were in a deep dilemma: if they left their homes to take shelter in Muslim refugee camps they would be killed on the way. But if they continued to stay in their own houses, they stood the risk of being attacked by Sikhs and Hindus. Their only hope was that the army would escort them safely to the refugee camps. Mridula worked amidst the abysmal conditions in these refugee camps. Hindus who stayed in Muslim majority localities of Jamma Masjid, Ajmeri Gate, Turkman Gate, Faiz Bazar, were equally afraid and the government deployed police and military to cordon off those areas. Despite all these measures, the government was not successful in bringing the situation under control. Faced with such a situation, Nehru had to request Gandhiji to reach Delhi from Calcutta at the earliest.

The Mahatma reached Delhi on the morning of 9th September 1947, where a grim-faced Vallabhbhai, the home minister, met him at Shahadara station. He was taken to Birla House, as *Bhangi* colony where he usually stayed, had been occupied by Hindu and Sikh refugees from West Punjab. In the car, Patel gave Gandhiji the facts of Delhi's lawless situation. Later that day, almost all the leaders including Jawaharlal met Gandhiji. Gandhiji assured the Muslims that no harm would come to them. So, many of them decided to stay on in their own houses. Elaborate arrangements were made for their protection and military and police pickets were provided in those areas where they were living. An Emergency Committee was formed and a dedicated team of officials and volunteers, men and women from several communities undertook the task of protecting Muslims, and organizing camps for Muslim, Hindu and Sikh refugees. Despite these measures, normalcy did not return and communal elements kept disturbing the peace and tranquility of the city. They also disrupted Gandhiji's prayer-meetings where he patiently and calmly preached unity and amity between Hindus and Muslims. On 13th January 1948 he went on a fast which was not to be broken till Delhi became peaceful.

To counteract these communal elements and restore peace, an organization by the name Shanti Dal (Peace Corps) was formed in Delhi by some members of the Indian National Congress, the Socialist Party, the Jamiat-ul-ulema and other progressive political parties. It did not have any regular membership but consisted of a group of dedicated workers who, were secular and non-communal in outlook. Mridula was elected its convener and was one of the active members. The first informal meeting of this group was held in her room in the Constitution House at New Delhi which also became the office of this organization. It was open almost round the clock and Mridula and her staff were always there to help people. The law and order situation in Delhi, which was rapidly deteriorating, was discussed at these meetings and it was resolved that full scale efforts should be made to mobilize the citizens to control and check the growth of communal feelings. During Gandhi's fast, a procession of Hindu, Sikh and Muslim women marched through Chandni Chowk and its lanes making a plea for peace. The Shanti Dal played an important role in restoring communal harmony and helped the police in maintaining law and order by forming vigilance committees in each ward of the city. Mridula put the Shanti Dal on a sound footing and brought it the patronage of the government. Everyday she sent a brief report of its work to Nehru. But she was primarily devoted to the work relating to the recovery of abducted women and children and had to remain out of Delhi most of the time, so Subhadra Joshi was appointed as a joint convenor.

At the request of the Shanti Dal, Jawaharlal attended one of its meetings, as did Maulana Abdul Kalam Azad, whose advice and guidance Mridula sought on issues regarding arrangements for the security and rehabilitation of the Muslims in Delhi. Due to efforts of Maulana Azad, Mridula and other members of the Shanti Dal, a meeting was held in the Ministry of Relief and Rehabilitation, where it was agreed that no Muslim occupying an evacuee-house was to be dislodged unless alternative accommodation was provided to him.

Hindu and Sikh refugees who had come to Delhi from Pakistan had occupied the land and property attached to mosques which had been abandoned. The Muslims in the walled-city were very sore about this, as they regarded it as a sacrilege and wanted the mosques cleared of the refugees. There were very few Hindus or Sikhs who sympathized with this demand, but among them was Mridula. She was, therefore, put on the Masjid Committee, which was asked to go into this question. Because of Maulana Azad's interest and intervention, Muslim residents of Delhi got most of their grievances redressed. Mridula also played an important role in getting concessions for them.

In August 1953, when Mridula started criticizing the Kashmir policy of the government of India and the arrest of Sheikh Abdullah, she was forced to resign from all Government related agencies including the Shanti Dal. Soon after her resignation in February 1954, the organization was wound up.

When the country was partitioned in 1947, large scale migration of population as in Punjab did not take place in East and West Bengal, partly due to Gandhiji's peace mission in Noakhali, Tripura and Calcutta. From the beginning of 1950, however, the Hindus in East Bengal were under increasing pressure to leave their homes and an exodus started, followed by Muslim migration from West Bengal. To check this, Nehru deputed Mridula to visit Calcutta and start relief work there under the aegis of the UCRW.

Soon after her arrival in Calcutta, she plunged herself into relief work and undertook a tour of the industrial areas of Hooghly, Howrah and Twenty-Four Parganas. She often faced stiff opposition from Hindu mobs who resented her efforts to rescue Muslims. She wrote about one such incident in 1951, prior to the first general election, where in the midst of arson, loot and murder, she faced an infuriated Hindu mob for three to four hours. Though the crowds were threatening, they were courteous. Some amongst them gradually saw reason and withdrew, while some hung on to protect her from their more infuriated colleagues. She was wondering how a group of beleaguered Muslims could be saved since the route was bursting with bullets and country-made bombs. As luck would have it, a military picket passed by, and in the twinkling of an eye, the rowdy elements disappeared and the road suddenly became deserted and desolate. Taking advantage of this, Mridula walked over to the Muslim side and found there a group of fear-stricken, entrapped people. She asked them to remain calm and not be afraid and stayed with them until the army arrived and took charge of the situation.[18]

Mridula then went to East Bengal to acquaint herself with conditions prevalent there. Both the Governor General, Gulam Mohammad, and the Prime Minister of Pakistan, Liaquat Ali Khan instructed the East Pakistan authorities to make all arrangements for her visit and show her around. She visited the Nawabganj camp in Dacca where fear-stricken and insecure Hindu refugees had taken shelter. She submitted detailed reports to the East Bengal government about these camps including suggestions on how they could be improved. Her findings were reported to Liaquat Ali Khan. She continued her efforts for the rehabilitation of unattached women and children in East and West Bengal but was at times deeply distressed and frustrated by the red-tape and delay in getting schemes through and of her inability to convince bureaucrats of the urgency and gravity of the problems. She pressurized the governments of Pakistan and East Bengal to start a home for unattached Hindu women and children and pursued the matter vigorously with Mr M.W. Abbasi, Secretary in the Ministry of Rehabilitation, Government of Pakistan and also with the Governor-General of Pakistan during her visit to Karachi in November 1950.

Durgabai Desmukh was appointed vice-chairman of the UCRW. After undertaking a tour of all the places where the productive centres of the UCRW were functioning, she submitted her proposal for the reorganization of the Council's work. The changes she proposed to bring about were not acceptable to Mridula, who felt that 'a terrible red-tape approach and wrong protocol notions' were creeping into UCRW. Durgabai, according to her, was overlooking the humanitarian aspect of the work, and also forgetting the urgent nature of their task.[19] The President, Dr Rajendra Prasad, informed her that the committee had full confidence in Durgabai and Mridula should not act independently or contrary to Durgabai's wishes[20] Since both Mridula and Durgabai were women with very positive views, they could not work together and Mridula withdrew from the UCRW. In any case, by this time, she had become increasingly involved in her work for the recovery of abducted women in Punjab.

CHAPTER EIGHT

RECOVERY OF ABDUCTED WOMEN

'*Bahenji*, our women and children are trapped on that side (Pakistan). Please get them out.'

'How did you manage to get here, leaving them behind?'

'What could we do? When we heard that trouble had started in our area, we came rushing out to look for help and then we could not go back.'

'...Now all of a sudden communications are broken off. Wild rumours are afloat about the terrible happenings. Please do something.'

'I have walked fifty miles. I swam across the river Ravi to tell our *hukumat* (government) that we are in great danger. Come to our help.'

From early morning till late at night, group after group of dazed and bewildered Hindu and Sikh refugees would flock to the Amritsar Hotel and demand help from Mridula. She listened to their tales with rapt attention and assured them that she would do her level best to get their women, who had been left behind at the mercy of the local Muslims, rescued at the earliest. Kali Prasad Dabral was in charge of all the administrative work in the Amritsar Hotel Office which was the headquarters of the Recovery Organization, at that time. He had worked with the Kasturba Gandhi Memorial Trust in Bombay before joining Mridulaben in Amritsar. Kamlaben Patel, a remarkably courageous and able Gandhian social worker was in overall charge of all the recovery camps.

Communal riots broke out in East and West Punjab on the eve of Partition and thousands of Hindus and Sikhs began to flee to India and likewise Muslims were crossing the border to seek shelter in Pakistan. On their way they were brutally attacked by huge mobs, armed with lethal weapons and women were abducted. The circumstances of abduction varied. Some had been separated from their families while escaping or strayed and were picked up. Some were left behind as hostages for the safe passage of their families; still others were given interim protection and then incorporated into the host family. There was hardly a family which did not have some near or distant women relatives abducted. Some women had put up fierce resistance against men who had tried to molest and rape them. Fear of abduction or of falling into the hands of the enemy compelled many to commit suicide. There were instances when women had jumped into wells and rivers in order to escape abduction and the loss of their chastity. In Thoa Khalsa, some ninety women jumped into a well. Infact the well became so full, that more women could not jump in. In Hanoli in Mianwala district, hundreds of women jumped into wells to escape being molested. Fathers killed their own daughters. Women jumped into burning houses and carried packets of poison to be swallowed in case they were captured.

Thousands of girls were abducted. Anees Kidwai in her extremely sensitive portrayal of these women in *Azadi ki Chhaon Mein* (Shadow of Independence) relates how the good *maal* (goods) could be shared among the police and army, the second rate stuff would go to everyone else. These girls would go from one hand to another and then another and then after several hours turn up in hotels to grace their decor, or would be handed over to police officers to please them.[1] A large number of abducted women were sold several times. Sometimes they were sent as gifts to friends and acquaintances.[2] There was the unforgettable train incident of 24 September 1947, involving refugees going from Tehsil Pind Dadan Khan, District Jhelum, towards East Punjab. About 3,000 male passengers were massacred and young girls were distributed amongst the police force, the national guards and the local *goondas*. The victims were collected in an open space and a free hand was given to the mob. 'After the massacre was over, the girls were distributed like sweets'[3] Those who were old were discarded and abandoned. There were hundreds of cases where families ran for their lives and the women and children were left behind and lost in the confusion.

The women suffered unimaginable cruelty and humiliation. For an abducted woman, there was nobody she could turn to for help. She had to live with a person who may have killed her husband, brother or father, her children or other members of her family. She was in a kind of prison house from which there was no escape. Her abductor believed that people of her community had killed members of his community and abducted their women. Hence he had no sense of guilt; in fact he took pride in having taken revenge by abducting her. Abduction as a retaliatory measure was both an assertion of identity and a humiliation of the rival community through the appropriation of its women. Deep down, women are treated as property. That is why such a fuss is made of 'our' women and 'their' women. As Levi Strauss says, the woman question is really a power question. A tribe asserts its dominance by 'taking' women from other tribes, but keeping its own women absolutely inviolate.

To rescue these helpless women from the clutches of their abductors, Mridula rushed to Punjab. Before anyone else's attention was drawn to the problem of recovering abducted women, she understood that it was not a question of nation, religion or community but of helpless women. Her task was to recover Muslim girls abducted and kept in Hindu and Sikh families and Hindu and Sikh girls abducted and held in Pakistan. She was convinced that this was humanitarian work of utmost importance and so devoted the next six years of her life entirely to it. Anees Kidwai writes that Mridula understood the gravity of the situation and being a woman of unusual courage, went to Pakistan and met officials in Lahore and Karachi.[4]

She set about the business of recovery of abducted women 'like a tigress with cubs', writes Gundevia who had been assigned the task of helping her by the Ministry of External Affairs. 'I had been warned', he writes, 'to be very careful (because) she had direct access to the Prime Minister'. If she had direct access to Prime Minister Nehru, she had also direct access to Prime Minister Liaqat Ali Khan. 'An abominable nuisance' was the common epithet used for her in Delhi and Lahore.[5]

The Indian official view shared by a large number of people was that abducted women must be restored. It was said that as descendents of Sri Ram, 'we have to bring back every Sita who is alive'. The Indian National Congress passed a Resolution at its Meerut session in 1946 (when Mridula was the General Secretary) that 'women who had been abducted and forcibly married must be restored to their houses.' This was reiterated by the Congress in November, 1947. Soon after Partition both the governments of India and Pakistan, as well as the governments of East and West Punjab declared that forced conversions and marriages would not be recognized and that women who had been abducted must be restored to their families and every effort must be made to trace and rescue such women. The Indian government mounted a massive 'rescue' operation which lasted for nearly nine years and aimed at recovering and bringing back abducted women to their 'rightful' homes, i.e. bringing Hindu and Sikh women to India and Muslim women to Pakistan. Both governments were agreed on this.

Apart from evacuating the refugees, the units of the Military Evacuation Organization were doing splendid work in rescuing abducted women and children. But the pace of recovery was slow. There were people who were prepared to rescue women if they were paid handsome amounts of money. Mridula rightly opposed this on the ground that women were not commodities to be purchased: 'this will ruin our work in the field of raising the status of women'.[6] She sent a note to this effect to Gandhiji, Nehru and Lady Mountbatten. It made a deep impact on them. The matter was discussed at an Emergency Cabinet Meeting where it was decided that the government was opposed to financial transactions of any form in the recovery of abducted women from Pakistan.[7] Gandhiji made a reference to this in a post-prayer speech, 'Some hoodlums come forward to bring back the girls if they are paid Rs 1,000 per girl. Has this thing become a business then?', he asked.[8] Soon after the prayer meeting, Mridula met Gandhiji and apprised him of the prevailing conditions in East and West Punjab. She related to him the episodes of abduction. He was deeply moved by the harrowing stories. He felt that every abduction should be regarded as illegal and it was the peremptory duty of both the governments to rescue these girls from captivity.

The recovery of abducted women was a gigantic task and Mridula did not have any prior experience of such work, nor was there any guidance available in the form of a precedent. She had plunged into the task guided simply by the instinct and desire of helping these women. She thought of various schemes and discussed them with politicians and social workers and also with Lady Mountbatten, who was extremely supportive. Mridula contacted leaders in India and Pakistan and impressed upon them the urgency of starting the recovery work. An Inter-Dominion Conference was held at Lahore on 6 December 1947, which unanimously agreed that the two countries should take steps for the recovery and restoration of abducted

women with Mridula as chief social worker. All abducted women
were to be restored to their respective Dominions even against their
own wishes. There was a debate on allowing women who declined
to leave their abductors, to stay where they were, but eventually the
views of Mridula and Begum Liaqat Ali Khan prevailed and it was
decided that they should be evacuated to the country where their
relatives had migrated. This was a complex question as it was not
always clear whether the women did not want to return because they
were afraid of saying so in front of their abductors or because of the
fear that they would not be accepted by their families since they had
slept with so many men or because they had settled down and did
not wish to be uprooted again. The Conference also decided that
statistics were to be compiled on standardized forms, giving par-
ticulars of women and children abducted in each Dominion, and that
rewards were to be given to those police officers who helped in their
recovery. The district liaison officers were to supply information
regarding abducted women. Their recovery was to be effected under
the joint police operations of East and West Punjab and women social
workers were to be associated with it. The local police were to be
assisted by a staff of the AIG, two DSPs, 15 inspectors, 10 sub-in-
spectors and 6 ASIs. Mridula strongly favoured the association of
women in the recovery work. Once when she was informed that the
DLO of Muzaffargarh district had left his district and it was not
known when he would return, she said the women workers on their
own should be allowed to do the rescue work.[9]

A joint appeal to the people of India and Pakistan to restore all
abducted women was made by the representatives of the Inter-
Dominion Conference. Prominent among the signatories were Ghaznafar Ali Khan, Iftikar Hussain Khan of Mamdot, Begum Liaquat Ali
Khan, Begum Shah Nawaz Khan, Begum Bashir Ahmed, Iftikharuddin, K.C. Neogy, Swaran Singh, Rameshwari Nehru, Mridula Sarabhai and Kamaladevi Chattopadhyaya

After the conference concluded, Mridula and Rameshwari Nehru
returned to Delhi, met Gandhiji on 7 December 1947, and submitted
to him separate reports. On that very day, addressing his prayer meeting, Gandhiji said about the number of Hindu, Sikh or Muslim women
abducted, which community did more evil, or who started abduction
was irrelevant. The important thing was how the evil was to be un-
done. The way to atone for this sin was to return the women. And

their families must take them back as they were not guilty and were pure.[10]

A separate section was formed by the Government of India, in the Ministry of Relief and Rehabilitation, to look after the welfare of women and children who had lost their guardians during the communal riots. Rameshwari Nehru was appointed Director and was also assigned the task of organizing the recovery of abducted women and children.

Since August 1947, Mridula's headquarters were at Amritsar and the first transit camp for recovered Muslim women and children was set up in Putlighar *mohalla*, outside the walled city by the side of the Grand Trunk Road.

Most of the abducted women were under 35 years of age and from villages. The moment a police vehicle would enter a village, the abductors would come to know of it and would run away and hide in a nearby field or forest area with the abducted women. As the police raids on these houses proved abortive, the police decided that the best time to search a house could be around sunset when all the family members were likely to be at home. The recovery police staff used to leave their vehicle and driver at the outskirts of the village lest the abductors got a signal about the police raid. There was a danger that if a girl was rescued from the village, the rest of the abducted girls may be sold or killed. To avoid detection, they walked down in twilight in separate groups of two or three each. Often the police party and the women social workers had to walk more than five kilometers. Generally, an 'informer' accompanied the police squad to lead them to the house of the abductor. The task was not easy as the abductor usually denied that the woman was abducted. Often, he offered stiff resistance and had to be physically overpowered by the police. Similarly, the abducted women had to be persuaded by the women social workers to accompany them to the government camps in a government vehicle as some of them were unwilling and others extremely frightened.

After recovery, the women were admitted to the Recovered Women's Camp which was guarded by the police. The condition of the camps was awful, worse, says Kamlaben Patel, than cattle sheds. They were overcrowded and because of lack of sanitary facilities, there were frequent outbreaks of epidemics and deaths. Within the limited budget, it was not possible to provide for more than two meals a day and a pair of clothes. The condition in some of the Pakistani camps was worse. When women from the Kurja camp arrived in Lahore, according to Kamlaben, they looked more like skeletons. They had not eaten properly for months, nor bathed for weeks. Their hair and bodies were covered with ulcers and lice. During the one hundred and ninety miles long journey from Kurja camp to Lahore, they had been provided no water to drink.

As news of the arrival of the inmates of Kurja camp reached East Punjab, relatives started coming in hordes in search of their lost wives, sisters and daughters. Security guards were placed at the gates as relatives tended to rush in at all hours. When Mridula brought Raja Gaznafar Ali Khan, Refugee and Rehabilitation Minister of Pakistan to the camp, the security guards stopped them as no one was permitted to enter without Kamlaben's permission. Mridula then pointed out to the Minister that 'maintaining discipline, law and order is not just the responsibility of men. You have seen today the ability of a Patel girl.' Gaznafar Ali Khan immediately sanctioned money for fruit, milk, etc. for the camp inmates.

The abductors seldom sat quiet. They tried all ways and means of getting back the women by going to the officers of the Recovery Organization and also to the camps where the recovered women were lodged, till they were informed that the women had revealed their true identity and it was fruitless for them to continue pursuing their case.

One woman social worker was attached to each police recovery squad in every district of East Punjab. The women social workers had to be extremely courageous and were recruited after an interview on the sole recommendations of Mridula. Many of them were political workers associated with the Congress, the Communist Pary, the *Kisan Sabha* or the Congress Socialist Party.

Many Hindus and Sikhs argued that if Pakistan did not recover Hindu/Sikh girls, India should also not help in the recovery of Muslim girls. But for Mridula, the recovery of a Muslim girl was as important as that of a Hindu girl. Her main concern was that the girl should be restored where she rightfully belonged. She held that there cannot be any bargain in the recovery work of Hindu and Muslim women. This kind of approach was resented by many Hindus and Sikhs but she had the support of Gandhi and Nehru which enabled her to face the opposition. She had no official designation nor did she ask for it. She was a mere social worker but because of her political contacts and status she had access to ministers and senior officials whom she could bully and order around. At the same time, she could mix with ordinary people also. While this made her popular among the refugees who relied on her for support, her method of work, impatience and desire for quick action often brought her into conflict with bureaucrats.

Soon after the Lahore Conference, a joint meeting of India and Pakistan was held at Jullundur on 22 January 1948. The home minister of East Punjab, Sardar Swaran Singh chaired the meeting which discussed, among other things, the recovery of abducted women. Mridula suggested that the police force sent by one province to the other should be increased and transport should be placed at the disposal of the special police staff by each province, to enable them to move out as soon as information was received. The time factor was important as their was little doubt that abducted women were being moved farther into the interior of both provinces. These suggestions were accepted. In addition to the transit camp at Amritsar, there were camps in Gurdaspur, Ferozpur, Ambala, Delhi, Jammu, Patiala and a few other places to lodge the recovered Muslim women and girls. The base camp at Jullundur was supervised by a Pakistani social worker under the deputy high commissioner of Pakistan at Jullundur. On a similar pattern, a base camp for recovered Hindu and Sikh girls was set up at Lahore in Sir Ganga Ram Hospital, which was supervized by Kamlaben Patel and Smt. Bhag Mehta, a Congress social worker. Transit camps for the recovered Hindu and Sikh women were also opened in the districts of Sheikhupura, Multan, Montgomery, Lyallpur and Sargodha. In each camp there were two Indian women social workers and a few Indian policemen. The social workers were recruited in Delhi by the Ministry of Relief and Rehabilitation, in consultation with Mridula. Many of them were fresh graduates from the Delhi School of Social Work and had no field experience. The recovered women were brought from the district camps to Lahore and kept in the Gangaram Hospital till there was a lorry full of them to send to Jullundur. The Gandhi Vanita Ashram Camp was started at Jullundur by the East Punjab Government to provide shelter to those 'unattached' Hindu and Sikh women and children who had lost their parents or guardians in West Punjab during the communal disturbances. It was also used to lodge the recovered women who were brought from Lahore. The director of the camp was Premavati Thapar and the person in charge of the camp was Krishna Kaushal.

Mridula contacted women of various shades of opinion and impressed upon them the necessity of giving wide publicity to the work being done for the recovery of abducted women. As a result, a joint appeal was issued on 28 January, 1948 by many leading women of India and Pakistan for expediting this work. A vigorous campaign was launched between 16 and 22 February 1948 which was observed as 'Restore the Women and Children Week'; Mridula made frantic efforts for its success. During this period the press ceased to be hostile and even the communal political parties became less critical.

Mridula visited Karachi twice when a rumour went around that abducted women were being removed from West Punjab and NWFP to Sind. She met the premier of Sind, Mr Khurro and requested him to issue the necessary orders for their recovery. She also met the Pir of Manki Sharif who promised her full cooperation for the NWFP.

Many Hindu/Sikh girls had been abducted when refugee trains at Jassarh (Sialkot district) and Gujrat (Punjab) had been attacked. The Pakistan government had prohibited the entry of Indian nationals to these districts as they had a common border with Jammu & Kashmir State. It also prohibited Indians from entering Rawalpindi, Jhelum and Campbellpur and had taken upon itself the responsibility for the recovery of the abducted women from those areas. Mridula was not satisfied with this arrangement and wanted to see the state of affairs for herself. She expressed her desire to visit these areas to Khan Qurban Ali Khan, and he agreed to arrange for her tour. Letters from Mridula to Pakistan officials regarding recovery of abducted women and children carried more weight than from other Indians because they knew that she was unbiased. As a result of her communication with the IG police, a number of Hindu and Sikh abducted girls were recovered from Sialkot, Gujrat and Jhelum and lodged in Kurja camp. She visited this camp and submitted a report to Raja Ghaznafar Ali Khan, Pandit Nehru and Lady Mountbatten.

When Mridula visited Kurja camp, she found a few women who were vehemently opposed to being taken to India, while some others were hesitating and wavering. She was not sure that they were telling the truth and wanted further investigation before any decision was taken. She firmly believed that every abducted woman wanted to return to her original family and home despite their protestations. She has described the Kurja camp incident in her own words.

'We know our relatives are dead. They were killed in our presence,' said some of the older women with tears in their eyes. A group of younger ones shouted hysterically: 'We changed our religion three years ago. Our whole family did so. Then the trouble started. The Maharaja's Dogra and Sikh troops raided and plundered our village and killed our relatives. We are Muslims; what right have you (addressing one Pakistani colleague) to hold us here and ask us to go with these *Kafirs*. If that is what you feel, why don't you go over to Hindustan yourself and leave us in peace.' The ring leader would then step forward, and the excited crowd would turn into a meeting. I felt as if I was being put on trial. On my capacity to handle their questions, on my tact and my ability to convince them lay the hope of making further progress.... A bright young girl took the floor and argued out their case. 'You say abduction is immoral and so you are trying to save us. Well, now it is too late. One marries only once—willingly or by force. We are now married. What are you going to do with us? Ask us to get married again? Is that not immoral? What happened to our relatives when we were abducted? Where were they? They now tell us they are eagerly waiting for us. No, you do not know our society. Life will be hell for us. Some of our nearest relatives are here, living as converts. We cannot leave them and go away. If you still insist, you may do your worst but remember this: you can kill us but we will not go.' She spoke with grace, dignity and determination. How I wished we could have her on our side, as a co-worker. About three weeks later, in the same camp, the same girl again took the lead and argued with a Pakistani official. 'Well take my case,' she said. 'Two relatives from Hindustan have failed to persuade me to return. They say my mother is coming this evening. That is not going to mend matters. I am your fellow-citizen of Pakistan. How dare you force me to go away against my wish? Those whose relatives have come to fetch them have gone. Now at least have mercy on us and let us be out of this camp.'

No heed was paid to her plea. Within a few hours, she was confronted with her mother who had been rushed from India. The change in her attitude was astounding and miraculous. Not only did she come over herself, but she brought the rest of the women round too along with her. That night she spent with her mother. Next morning, I spoke to her in the presence of the Pakistani officers. She was too ashamed of herself to look at me. Shyly but with a radiant expression, she exclaimed, 'So, it is you who have been after me.' We burst out into a hearty laugh.

She asked the Pakistani officers to forgive her and added, 'I have no relative here. Nor have the other women I know here. We are all mortally afraid, and hence gave you this trouble. Here are the names of some of them. Tell them everything is OK in Hindustan and they are to come over. They will come.' Later in the day, I asked her to use her influence with another girl, also a very difficult case. In her arguments with this other girl, we heard her saying, 'Until yesterday I too was like you. But you will not be able to face your mother if she were to come here. And that *Behenji*, with the other *Behenjis*, will spare no effort to induce you to come away. Why then do you want to go through the same futile agony as we have foolishly gone through.' she is now helping us in recovering her friends and is also taking a workers' training course. Thanks to the all-out efforts of the workers, the press and the government agencies, within ten days we were able track down the relatives of this group of women and rush them from India to Pakistan.[11]

Kamlaben Patel relates several such incidents. In the Shekhupura district camp in Pakistan there was a 20 or 21 year old girl who claimed that the police had forcibly brought her and that she had fallen in love with a Muslim boy whom she had willingly married, hence hers was not a case of abduction. A few days later, a young, Hindu man came to the camp looking for his wife and on seeing that same girl in the camp, recognized her as his wife. Confronted by her husband, the girl fell sobbing into his arms and confessed that she had been lying out of fear. She and her husband had been caught while trying to escape from Pakistan and as he had been severely beaten up, she thought he had been killed. She was sold for fifty rupees. She had no relatives in India and as the Muslim boy who bought her was kind to her and treated her well, she had made up her story.

Similarly, Veera, a Hindu girl, had been given by her father to a Muslim Sub-Inspector of Police in Multan in return for which her family was promised safe passage to India. Brought to the camp, Veera claimed that she had willingly married the Sub-Inspector and she had no wish to return to India. 'Are you not ashamed,' she said, 'to forcibly drag a married woman like this into the camp? I want to stay with my husband.'

Then one night, she went to Kamlaben and asked her: 'Everyone in the camp says that there is a lady who wears salwar khameez, has bobbed hair and is so powerful that she can get anything done in India. Pandit Jawaharlal Nehru even listens to her. If I tell the truth, will she help me?' And then she came out with her true story. The fact that her parents bought their safety at her cost, hurt her deeply and hence she did not want to go back to them. However, she agreed to go back to India, as it was decided that she would not be forced to stay with her family.

Mridula wanted to rescue the women who had been abducted in Pakistan occupied Kashmir and for this purpose went to Rawalpindi and met Sardar Mohammed Ibrahim, the President of '*Azad*' Kashmir government. They discussed at length the recovery and restoration of abducted women belonging to Jammu and Kashmir State. Sardar Ibrahim agreed to make all efforts to recover abducted women, provided the government of India and Jammu and Kashmir administration did the same in their territory. An agreement was signed between Sardar Ibrahim and Mridula on 7th March 1948 which was a great achievement and the initiative taken by her was widely appreciated.

The recovery organization in Pakistan relied a great deal on Mridula for support. Rabia Sultan Qari, an eminent social worker of Pakistan, who was engaged in the recovery work, undertook a tour of Patiala, Jind, Alwar and Kapurthala and submitted a report to Mridula on the conditions in the abducted women's transit camps which she found to be unsatisfactory. Miss Qari's report was forwarded by Mridula to her Highness, the Maharani of Patiala who promised to take action to improve arrangements in the camps.

The question of abducted women was discussed once again at the Inter-Dominion meeting in May 1948. It was broadly agreed that both the countries would cooperate with all available publicity media to rouse public conscience against the keeping of abducted women. It was estimated that the total number of abducted women in Pakistan was over 12,500, but Mridula's estimate was much higher.

Another Inter-Dominion Conference was held at Lahore on 5th July 1948 to iron out the differences between India and Pakistan and to step up the recovery work. But throughout the Conference, charges and counter-charges were levelled by the representatives of both countries against each other. Sharp differences arose between Rameshwari Nehru and Mridula. The former held that a firm stand should be taken with the Pakistan government. Mridula disagreed and opined that their approach should not be retaliatory but conciliatory.

She wrote that India's case was weak in the Conference since in case of Joint Enquiry, officers in charge of the work in East Punjab were extremely dilatory and despite repeated reminders did not move fast. While Pakistan had handed over a few women to their converted relatives without India's approval it was revealed at the Conference that India had done the same in Jullundur according to Rameshwari Nehru, under the orders of the Government of India. She wrote that she was shocked and greatly hurt to hear this because she felt that by this act she was let down by her colleagues in charge. She had strong views on the subject and felt that she ought to have been consulted.[12] She believed that the recovery of abducted Muslim women should proceed in India, irrespective of the number of Hindu and Sikh women recovered from Pakistan. Once the Ministry of Relief and Rehabilitation and its Women's Section under Rameshwari Nehru became well established, they started to direct the recovery work, often bypassing Mridula. Till December 1949, the number of recoveries from India was 12,000 and only 6,000 from Pakistan. Since the number of Hindu and Sikh women was less, Rameshwari Nehru felt that the discontent among Hindus and Sikhs in this regard was genuine and could not be ignored.

On thinking over the whole issue, Mridula concluded that she should be given a free hand or she should withdraw from the organization. She wrote a letter to this effect to Gopalaswamy Iyengar, Minister Without Portfolio, who had been assigned the work relating to the recovery of abducted women by the Prime Minister.[13]

In the beginning, Mridula had been given considerable freedom but the officials now resented a 'non-official' functioning with full responsibility and authority, and wanted the entire operation brought under their control. Under the circumstances, she felt that she could not serve any useful purpose. For Mridula the recovery work was not only a humanitarian problem but a part of her ideology.[14] She gave in her resignation but Nehru, instead of accepting it, put her solely in charge of the recovery work, transferring it from the Ministry of Relief and Rehabilitation and bringing it under the Ministry of External Affairs. Mridula was provided with a new office, the Central Recovery Office, which was set up in Constitution House, New Delhi, where she stayed in those days. The Central Recovery Office kept in close liaison with the Recovery Organizations in India and Pakistan and helped them in the implementation of government policy regarding the recovery of abducted women and children. Four months after this, a special ordinance was passed vesting in the hands of the police the power to recover. Though a feminist in many respects, Mridula did not realize that in this case she was not giving women the right to decide their fate.

A permit system was introduced between the two Dominion governments from 19 July 1948 and this slowed down the recovery work. This caused considerable worry to Mridula and she moved the Government of India to take up the matter at the highest level with the Pakistan government. She repeatedly made the point that Indian officials should not adopt a retaliatory attitude. If Pakistan delayed or adopted an obstructive attitude, India should not follow suit. Recovery had to be carried out on its own merit irrespective of what the other government did.

She went to Lahore to discuss with Pakistan officials and social workers the extremely slow progress of recovery work there since July 1948 and ways of expediting it. She met Mohammed Yar Khan, Under Secretary, N.A. Rizvi, Superintendent of Police, Khan Quarban Ali Khan, I.G. Police and Begum Fatima, a social worker. She sought an interview with Khwaja Shabuddin, Minister for Refugees in Pakistan and found that he had been given a highly exaggerated account of some incidents in which Begum Fatima had been ill-treated at the Indian check-post. It took some time for her to convince him that he had not been briefed properly. She explained to him that Indian women social workers were also often ill-treated in Pakistan but instead of complaining to officials in India, they tried to sort out their problems with Pakistan officials. She pleaded with him that the recovery of abducted women should be treated as humanitarian work and kept aside from the Inter-Dominion controversies. Khwaja Shabuddin assured her of his government's determination to carry on the work. With the permission of the West Punjab government, she visited a refugee camp in Lahore where a group of Hindu and Sikh women recovered from Pakistan as well as Muslim women recovered from India were kept together and talked to them.

From Lahore, Mridula flew to Karachi and had an informal meeting with the Finance Minister, Mr Ghulam Ahmed at his house. She also requested the Minister without Portfolio, Mr Gurmani to take steps to improve the condition of Kashmiri Hindu girls in the Lahore camp and proposed that these girls should be separated from the Muslim girls for psychological reasons. Mr Gurmani requested her to obtain authority from the Government of India, confirming that she was their accredited representative for conducting negotiations with the Government of Pakistan. Mridula wrote back that she was serving the Recovery Organization in the capacity of a social worker for the sole purpose of mitigating the sufferings of abducted women. She wanted to keep herself above the prejudices of the two governments and examine each problem from a purely humanitarian angle.

It was because of this approach that she was able to win the cooperation and goodwill of the Pakistan government and was even requested to provide guidance to Muslim social workers who were in charge of refugee camps. She was trusted by Pakistani officials and had access to all top level politicians and bureaucrats. She could visit any remote part of Pakistan alone even where communal riots were raging. This was indeed a unique tribute, when relations between the two countries were so bitter. Mridula had crossed the Indo-Pakistan border several times and goaded Liaquat Ali Khan and several others in Pakistan into agreeing that some proper machinery should be set up for the recovery of abducted women. Women social workers had to be organized, the police had to be given special powers etc. 'If India did all this, would Pakistan do the same?' Nobody believed that Pakistan would play ball (except Mridula). She placed a fourteen paged document in the hands of Gopalaswamy Iyengar and told him that she had discussed this with Liaquat Ali Khan and Pakistan officials. 'If you sign this,' said Mridula to Gopalaswamy, 'I can get Liaquat Ali to sign it tomorrow.' The secret of her hold over Liaquat Ali has been revealed by Gundevia, a joint secretary in the External Affairs Ministry who was then dealing with the recovery work.[15]

Begum Liaquat Ali Khan was originally a Hindu who had first been converted to Christianity and then to Islam. During the ghastly communal riots, she and her husband had given refuge in their house to a little Hindu girl, orphaned for all they knew, who had been brought to them by somebody. 'Liaquat had given Mridula the name of her parents. If she ever traced the parents, he wanted the child sent back to them. If she could not find the parents he had meant to keep the child, safely, in his house for her life as a living memory of the tragedy.' Mridula hunted for the parents and, believe it or not, found them. The child was quietly restored to her parents, without anyone knowing about it.[16]

Gopalaswamy gave the fourteen-page document to Gundevia, and asked him to scrutinize it and put it into some reasonable shape. Gundevia then called Shrinagesh, Commissioner of Jullundur division with whom Mridula had a good equation. With great difficulty, Mridula was kept out of the room and Gundevia and Shrinagesh drafted a new four-page agreement and took it to Gopalaswamy for his approval. But how were they to persuade Pakistan to sign it? 'Simple,' said Mridula. She, Gundevia and Shrinagesh should fly to

Lahore. 'She would see that the Chief Secretary, West Punjab, Liaquat's Principal Secretary and others would meet us there; and we would settle everything with them. Mridula had already posted her fourteen pages to Liaquat 'so they know what we are after', she said quite calmly.'[17]

Subimal Dutt, who was then Foreign Secretary, was appalled. This was no way to conduct an international agreement and at a joint secretary's level at that. But finally after a great deal of discussion between the Minister, the Commonwealth Secretary and the Foreign Office, the agreement was cleared and Mridula was asked to arrange a meeting in Lahore. The Ministry of External Affairs telegraphed Y.K. Puri, an ICS officer, who was India's deputy high commissioner in Lahore. Mridula shot off her own telegrams to Liaquat Ali Khan in Karachi and to half a dozen other officers in Lahore. Less than four days after Gopalaswamy had given the green signal, Gundevia, Shrinagesh and Mridula were in Lahore. Mridula had gone ahead of them. She was completely acceptable to Pakistani officials—because of the threat of her being 'very dangerous' and having 'direct access' to their prime minister. Thus a meeting of the representatives of India and Pakistan was held at Lahore on 11th November 1948 virtually entirely due to Mridula's initiative and effort. After considerable discussion a draft agreement was prepared. The Pakistani officials said that their prime minister would have to approve the agreement they had hammered out. Mridula who was in the room, said, 'Oh, Liaquat Saheb will definitely approve'. One of the officials, obviously angered, said, '*Bahenji*, you seem to know more than we do, don't you?' Writes Gundevia, 'Mridula was puffing like an angry bull, nostrils dilated. But the Pakistan Chief Secretary smiled and said something nice about how helpful *Bahenji* had always been, and *Bahenji* was just a little flattered, cooled down, and to my surprise, kept her peace.'[18] It was agreed that the responsibility of recovery of abducted women would henceforth be that of the police and civil authorities of the respective Dominion in which the abducted women were residing. The recovered Hindu and Sikh women from Pakistan were to be handed over to the transit camp at Lahore which would be under the control of two Indian women social workers. Recovered Muslim women from India would be lodged in the transit camp at Jullundur under the control of two Pakistani women social workers. Any difficulty or legal complications arising in the process of recovery was to be enquired into and resolved by the superintendent of police and the case under dispute was to be referred to high powered officers of the two countries for resolution.

In accordance with this Agreement, the sole responsibility for recovering abducted Muslim women from East Punjab was that of the Government of India and the entire official machinery engaged in this task was geared up by Mridula. The recovery of Hindu and Sikh women also now began to proceed more satisfactorily in Pakistan.

Mridula raised a heroic team of women volunteers who did more than half the work, leaving only the difficult cases to the police. There were no complaints of any force being used in the recovery work. As soon as the cases of the abducted women were decided, immediate arrangements were made to transfer them from Lahore to Amritsar and vice-versa. They were accompanied by a police squad and women social workers and were not subjected to any search by the customs authorities at the check-posts. The news about the arrival of the recovered women always reached in advance and their relatives were usually anxiously waiting for them.

A Bill for the Recovery and Restoration of Abducted Persons was introduced in Parliament by Gopalaswamy Iyengar which became an Act in December 1949. Clauses 1 and 2 of the Act pointed out that: (1) Every effort must be made to recover and restore abducted women and children within the shortest possible time. (2) Conversion by persons abducted after March, 1947 will not be recognized and all such persons must be restored to their respective dominions. One of the principal features of this Act was that it adopted a more comprehensive definition of the term 'abducted' than the one provided in the Indian Penal Code. Another important clause was the setting up of an Indo-Pakistan Tribunal to decide the disputed cases of abducted women. According to the provisions of this Act, no option was given to the recovered persons; and they had to be sent to the country where their relatives had gone. The wishes of the persons concerned were irrelevant and consequently no statement of such persons were to be recorded before magistrates. Mridula was the driving force behind all this. She explained how every technique was used by abductors for defeating the recovery efforts and this legislation was necessary because it made abduction of women and children a crime, liable to severe punishment. Mridula held the view that India should go ahead with the work even if Pakistan did not respond adequately. She pointed out the enormous difficulties involved in tracing the abducted women and then recovering them. She could not understand the opposition to recovery work, which was essentially an effort 'to remove from the lives of thousands of innocent women the misery that is their lot today and to restore them to their legitimate environment, where they could spend the rest of their lives with *izzat* (honour)'.

When the debate on all the clauses of the bill was over, and some of the officials had adjourned to Gopalaswamy Iyengar's room in Parliament House,

a very excited Mridula said to the Minister, 'Thank God, Sir, it is all over and the women in both countries are going to be grateful to you.' 'It is not over, said the Minister. 'One very important matter still remains—I am writing to your father to have you abducted from Delhi and tucked away in Ahmedabad for some time,' he said with that genuine sense of humour which was so very characteristic of him.[19]

The Government of India's Ordinance as well as the Act regarding recovery of abducted women greatly facilitated recovery work of abducted Muslim women and children. On the basis of reciprocity, the Pakistan government also promulgated an Ordinance followed by an Act thereby disappointing Mridula's critics who had said that India was taking a big risk.

In early 1952 the East Punjab High Court gave a judgement that no recovered Muslim woman could be repatriated to Pakistan as she had become a citizen of India after the promulgation of the Constitution of India. Because of this judgement, the recovery work came to stand-still in East Punjab and other parts of India. This judgement was however challenged by the Government of India in the Supreme Court. Till the announcement of the verdict of the Supreme Court, Mridula continued the recovery work in Jammu and Kashmir as she knew that the orders of the East Punjab High Court were not applicable in that State as it had its own separate judiciary. The relatives of these Muslim women recovered from Jammu were either residing in Kashmir valley or had migrated to Pakistan. The only possibility of restoring the recovered Muslim women whose relatives were in Pakistan was through the Cease-Fire Line. Mridula, therefore, approached the Ministry of Defence, Ministry of Home Affairs and the Ministry of External Affairs of the Government of India and obtained their approval to select a place on the Cease-Fire Line Zone where the Indio-Pak Tribunal of the Superintendents of Police for

Recovery could hold their meetings. After consulting the government of Jammu and Kashmir State, Suchetgarh border, which lay between the districts of Jammu and Sialkot, was selected. The approval of the Pakistan government was also obtained for holding the meetings at the Suchetgarh border. The meetings were held there every month. The recovered Muslim women were taken to the border where their relatives from Pakistan and India were also brought. The Indo-Pak Tribunal usually decided the case in favour of those relatives who were closer to the recovered women. It was a great achievement on the part of Mridula to get the Cease-Fire Line opened in the interest of the recovery work. Subsequently, the Supreme Court of India over-ruled the judgement of the East Punjab High Court and the work of recovery of abducted women started again in East Punjab and other places.

Pakistan invaded the state of Jammu and Kashmir on October 26, 1947 with the assistance of the tribal hordes. The military and the police force of Maharaja Hari Singh, ruler of the state could not stop them. The raiders killed the Hindus and Sikhs of Muzaffarabad, Mirpur, Poonch, Rajauri and other areas, and looted their properties. A large number of Hindu and Sikh girls were abducted, raped and sold freely to Muslims. Mridula came to know that more than one thousand girls were in the possession of these Muslim abductors. Besides, some male survivors of these areas who had escaped assassination were imprisoned in a camp by the authorities of Pakistan occupied Kashmir. They were sending frantic messages to the Deputy High Commission of India in Lahore for their early evacuation. Mridula took up the matter with the government of West Punjab and through their good office was able to contact Sardar Mohammed Ibrahim, the Supreme Head of the 'Azad' Kashmir government and Chaudhary Ghulam Abbas, President of the Muslim Conference of Jammu and Kashmir state who had migrated to Pakistan.

Chaudhary Ghulam Abas' daughter had been abducted by an Indian military officer at Satwari near the city of Jammu and during the course of his talks with Mridula, Sardar Mohammed Ibrahim made a special request for the immediate recovery of this girl. Mridula promised to do so. She deputed a police officer specially for this work and was ultimately successful in getting her recovered after a long time. As a result of her sole efforts, she was able to get one thousand and two hundred Hindu and Sikh women and children recovered from the 'Azad' Kashmir Zone and also got evacuated the stranded male survivors.

On 1 July 1953, Mridula met General Ayub Khan, Commander-in-Chief of the Pakistan army, and requested him to issue orders for the release of the abducted Hindu and Sikh girls who were in the possession of the Pakistan armed forces. He assured her that he would do so.

Recovery work began to slow down by 1952–3 as the abducted women seemed to be settling down. As time went on, the abducted women began to be assimilated in the family and society of their abductors. The women gave birth to children, were treated not as 'keeps' but as wives and adjusted to their new way of life. Many abductors filed suits in lower courts against the recovery police officers and the camp-in-charges to get back the recovered women. After the Constitution of India came into being, *habeas corpus* petitions were also filed. In the first of such cases, the Punjab High Court ordered the release on bail of an abducted girl who was in a recovery camp. Mridula was greatly upset and submitted a note to Jawaharlal Nehru giving the entire background. There were several tragic cases: Boota Singh, a fifty-five year old Sikh bachelor farmer, had purchased a seventeen year old Muslim girl, Zainab from her abductor for Rs 1,500 in 1947. He married her and they had a daughter. A nephew of his, having an eye on his property, reported the presence of Zainab to the authorities. Mridula's co-workers recovered her, kept her in a camp for six months and finally sent her back to Pakistan. Boota Singh tried all means of getting back his wife. He became a Muslim, went to Lahore with his daughter Tanveer and on being rejected by the relatives of Zainab, in despair, committed suicide. Jitu, a Hindu boy and Ismat, a Muslim girl were in love in Lahore. After partition, Jitu's family left for India. Ismat managed to escape from Lahore with Kamlaben's help and married Jitu in Amritsar. Her family lodged a complaint of abduction and she was forcibly sent back to Lahore where her family pressurized her into denying all relationship with Jitu. A heart broken Jitu returned to India. Sudarshan a Hindu girl, studying in Lahore was in love with a Muslim zamindar's son in her college. Sudarshan's family migrated to Delhi after partition, and the boy followed her and they married and left for Lahore. Soon after, her father and brother came to Lahore and demanded that she be restored to them as she had been abducted, but Kamlaben helped Sudarshan to go back to the man of her choice.[20]

As years passed, Rameshwari Nehru stressed the human angle and said that while figures of recovery were encouraging, figures were not the only criterion for the work. She wondered whether it was humane to uproot these women and their children and she was against women being forcibly removed from the new families in which they had settled down. She had received a report of some Muslim women who resorted to hunger strike in Ambala as a protest against the attempt to forcibly send them away.[21] The Commissioner of Jullundur Division, A.L. Fletcher also shared this view. He interviewed an abducted woman, Ahmed-Un-Nissa and was convinced that she was happy and did not want to return to Pakistan, and so was against compelling her to do so.

The case of Ahmed-Un-Nissa became more complicated as while her abductor filed a suit in the Punjab High Court against her detention the Government of Pakistan brought increasing pressure on the Government of India for her early transfer to Pakistan. Her parents and relatives were also pressing for her early repatriation and Ghulam Haider, Superintendent of Police in charge of recovery of abducted women in Pakistan, wrote to Mridula that he met Ahmed-Un-Nissa many times and felt that her case had to be tactfully dealt with. The abductor also met her to expedite the decision. Mridula wanted the girl to be taken to Pakistan and given an opportunity to meet her relatives and stay with them for some time before deciding whether she wanted to stay on or return. Fletcher had strong objections to this procedure and wanted her wish to be ascertained before she was sent across. Mridula felt that these abducted women suffered from an acute fear complex and would not speak openly and frankly until they felt safe and secure. According to her, the abductors frightened the women who were told that they would not be accepted by their families and would even be murdered. All their relatives had been killed, so where would they go? As the women had no opportunity of meeting anyone outside the homes of their abductors, so it was difficult to remove their misapprehension and fears unless they were rescued.

Because of his differences with Mridula, Fletcher wanted to resign as High Powered Officer. But she wrote him a very polite reply saying that there was no fundamental difference in their approach and urged him not to resign. Mridula agreed that the recovered woman's consent was now necessary, if she was an adult, but argued that the consent should be taken after her fear complex was removed and she had spent some time with her relatives. Mridula felt that Muslim women would be better able to make up their minds after being taken to Pakistan, and Hindu and Sikh women after being brought to India. Despite her secular approach, Mridula could not accept that an abducted Hindu or Sikh girl could be happily married to a Muslim who had forcibly abducted her or vice versa. Rameshwari Nehru said that many Muslim women who had been forcibly sent to Pakistan could not be restored to their families and were given away in marriage to strangers without their consent. 'By sending them away we have brought about grief and distortion of their accepted family life without in the least promoting their happiness.'[22]

Cases were often not decided for months. In the Jullundur camp there was a Muslim girl from Alwar all of whose relatives had died in the communal riots. She had been in the camp for six months and was pregnant. When the Indian police brought some distant relatives of her's from Alwar, Pakistan officials alleged that these were not her real relatives and she would be sold by them. And so her case went on and on. The unfortunate girl when asked whether she wanted to remain in India or go to Pakistan replied: 'I cannot make up my mind. I am all confused. I will do what you tell me.'

Since the recovery process could take several years, by the time the women were found, they were often married, settled, with children, perhaps living a new, a different life. Some had been through so many hands that they had no longer faith or belief that just because their rescuers were from 'their' religion, they were trustworthy. Sadat Hasan Manto's *Khol Do* poignantly describes how a young woman was rescued and repeatedly raped by her rescuers. Several abducted women did not want to return, particularly Hindu and Sikh women, who feared that their families would not accept them. Their fear was not entirely unfounded as many families did refuse to take them back and Gandhi, Nehru and others had to issue appeals to people to do so saying that these women were not sinful. Mridula, however, was convinced of the righteousness of her mission and refused to believe that an abducted woman did not want to go back to her original home. We do not find in her papers the moving stories and poignant accounts of the suffering, dilemmas and problems of abducted women that we come across in Kamlaben's book. Of course, there were women who refused to accept their forceful marriages to their abductors and were happy to return to their families but there were others, who were afraid of being uprooted once again.

Anees Kidwai describes the plight of some of those girls and how in many cases when in the midst of murders, killings, loot and rape, a man picked up a woman and took her home, 'she looked upon this man as having rescued her, saved her from further public humiliation and she would forget her mother's slit throat, her father's bloody body, her husband's trembling corpse and look upon this man with gratitude. Rescuing her from beasts, this man had at least brought her to his home, given her respect and offered to marry her. After all she had been through, she was afraid to face her parents or her husband, if they were alive.'

On 9 August 1953, Sheikh Abdullah, then the prime minister of Jammu and Kashmir state was arrested. Miridula's open support for Sheikh Abdullah and propaganda against Bakshi Ghulam Muhammad gave an opportunity to her opponents and some central ministers to put pressure on Nehru to remove her from the Central Recovery Organization. By this time, her approach to the recovery programme was being questioned by many. So, she proceeded on leave and handed over charge. Her letter offering her resignation was extremely polite and there was no bitterness or anger in it—remarkable, considering that she had, day and night, for the past six years, devoted herself to this task. Her disassociation from the recovery work was a shock to many of her colleagues who felt that a vacuum had been created, that the humanitarian aspect would be lost and the work would be sabotaged. One of them wrote:

I was a little horrified to hear from Delhi that you are no longer working Recoveries. I suppose the show will gradually slow down and die, without you.... Your tremendous energies should be diverted to something more profitable. I know you have a one-track mind on this issue, but I hope you are now allowing yourself time to look at the other point of view. You know how completely I was in agreement with your views on the subject, at least three years ago. But these three years, and every single day of these thousand days have made a difference to each individual case that you and I were after. Look at it that way, and I hope it will console you.[22]

1. A view of 'Retreat' Bungalow from 1911 to 1935 which was thereafter renovated as per Surandranath Kar's design.

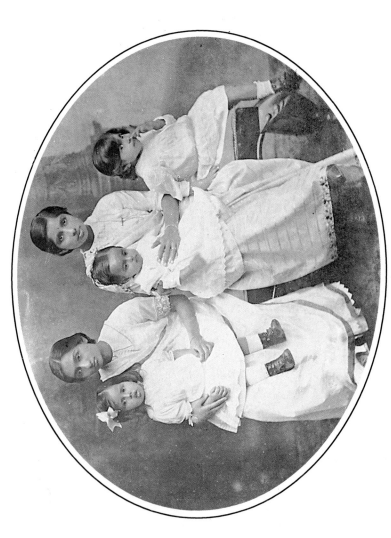

2. Mridulaben (extreme right) holding onto her mother Sarladevi (centre) with third baby son Suhrid in her lap and sister-in-law Nirmalaben (right) with Bharatiben in lap—1915.

3. Mridulaben and Bharatiben with an English governess—1915.

4. Mridulaben with her parents, three brothers and four sisters—1925.

5. Mridulaben horse riding in Ahmedabad—1921.

6. The Sarabhai family in Shillong with Subhash Chandra Bose and his brother Sarat Chandra Bose—1927.

7. Mridulaben, Secretary (Women's Sub-Committee) in National Planning Committee with Jawaharlal (Chairman) and G.P. Hutheesing (Jt. Secretary). Bombay—1940.

8. Mridulaben with Gandhiji in Bihar—1947.

Though she disassociated herself from the Recovery Organization in an official capacity, she continued to appear as a witness and fight out the cases of Muslim recovered women in the East Punjab High Court. People were afraid to appear as witnesses where Muslim recovered women were concerned so the Muslims largely depended on her. Most of the cases were filed by abductors with the backing of Hindu and Sikh communal parties.

The work of recovering abducted women and children came to a grinding halt towards the end of 1953, though the Abducted Persons (Recovery and Restoration) Act continued to be renewed every year upto 1957. According to Anees Kidwai, by 1954, 17,000 women were rescued from Pakistan and 20,000 from India. By then, the abducted women began to show increasing reluctance to go to the other country, leaving their children behind. By the Indo-Pakistan government decision of 1954, they could not be forced to go against their wishes. Moreover, the most serious consideration which prevented the Government of India from renewing the Abducted Persons (Recovery and Restoration) Act of 1949 was the problem of post-abduction children. The Government of India had passed an Ordinance that women whose babies were born in Pakistan after partition would have to leave them behind since it was felt that in Hindu society, a child born to a Hindu mother by a Muslim father would not be accepted. Pakistan also had a similar rule. Many social workers opposed such a callous solution to the problem which forcibly separated infants from their mothers. Kamlaben could not agree with Mridula Sarabhai's 'practical' and 'unemotional' approach. Between 1st January 1954 and 30th September 1957, no fewer than 860 children were left behind by Muslim women who went away to Pakistan, whereas 410 children were taken by them. The children left had to be taken care of by the government and put in orphanages. The administration gradually wound up the Recovery Organization district by district. Mridula also moved out of this work and for the next twenty years her sole preoccupation was Kashmir.

While many politicians and public figures distanced themselves from Mridula because of her involvement in the Kashmir affair, ordinary people such as taxi drivers of the stand near Constitution House in New Delhi, where she had her office for the Recovery of Abducted Women and Sikh taxi drivers even in Bombay, railway officers, security guards in North Block and South Block secretariat in New Delhi and people in other unexpected places, even years later, remembered her work, would recognize and help Kamlaben because of their gratitude to Mridula and her co-workers for rescuing some relative of theirs from abductors in Pakistan. Everywhere it was the same refrain: 'You are our *Behenji* who rescued my wife, or sister, or daughter, it is our pleasure and duty to help you... how can we accept any money from you when you did so much for us.'[24]

CHAPTER NINE

KASHMIR

Accession—To Be or Not to Be

On 9 August 1953 at 4.30 a.m. a notification was issued by *Sadr-e-Riyasat* of the Kashmir State, Yuvaraj Karansingh dismissing Sheikh Abdullah from the Prime Ministership and dissolving his Council of Ministers. Sheikh Saheb was arrested and was accused of treasonable correspondence with foreign powers. He was lodged in Tara Niwas in Udampur while Mirza Beg, his close colleague, was taken to the Central jail in Jammu.

In order to grasp the significance and importance of the sudden turn of events in Kashmir one has to try and understand the strategic importance and the political history of the state.

Jammu-Kashmir state is in the north-western part of India and has a common border with four countries—the USSR, China, Afghanistan and Pakistan. The heart of the state is the beautiful valley of Kashmir through which flows the river Jhelum. To the north of the Valley is the rugged and rocky upland of Gilgit and Hunja; and to the south is the province of Jammu. Towards the western side of the Valley lies Muzaffarabad, Mirpur, Poonch and Reasi and to the east is the province of Ladakh. At present, Gilgiht, Muzaffarabad and Mirpur are beyond the Cease Fire Line and are under the control of Pakistan. The rest of the state is in India.

At one time Kashmir was a great seat of Hindu and Buddhist culture. The conversion of Hindus to Islam took place during the reigns of the Muslim rulers from the fourteenth century onwards. Today, the population of Muslims in the Valley is over 90 per cent while Jammu has a majority of Hindus and Ladakh has a Buddhist majority.

In 1947, at the time of the transfer of power from the British government to the Indian government, Maharaja Hari Singh ruled the Jammu and Kashmir state. He was a descendant of Maharaja Gulab

Singh who belonged to the ruling clan of Dogra Rajputs of Jammu. Gulab Singh had joined the army of the Sikh Maharaja Ranjit Singh and in appreciation of his valour and loyalty, he was appointed Raja of Jammu. In 1846, Gulab Singh acquired the Raj of Kashmir through a treaty with the British and the descendants of Ranjit Singh.

Sheikh Mohammed Abdullah was born on 5th December, 1905 and belonged to a middle class Kashmiri Muslim family. After obtaining a M.Sc. degree in chemistry from the University of Aligarh, he served as a teacher in the Government High School at Srinagar but because of his political activities, he was dismissed from service. But this did not crush the Sheikh. With the support of a few friends and well-wishers he raised a political party in 1930 and named it Muslim Conference. The party quickly grew in strength and within a few years, became representative of the aspirations of the Muslim masses of the state.

Sheikh Abdullah first met Jawaharlal Nehru in 1937 at Lahore when the latter was proceeding to the North West Frontier Province (NWFP) at the invitation of Khan Abdul Ghaffar Khan, popularly known as Badshah Khan or Frontier Gandhi. He accompanied Nehru to NWFP and was introduced to Badshah Khan. This meeting was a turning point in his life as it brought a radical change in his political thinking and he began to view life with a broader outlook. On returning to the state, he, with his followers, converted the Muslim Conference to the National Conference in 1939 and opened the new party's membership to all communities. Since then, he associated himself with the Indian National Congress and attended its annual session at Tripuri the same year, where he came in contact with many Congress leaders and prominent workers, Mridula being one of them.

Sheikh Abdullah became a vehement critic of the two-nation theory propounded by Mohammed Ali Jinnah and an ardent believer in Hindu-Muslim unity as preached by Mahatma Gandhi. This helped him to secure moral and material support form the Indian National Congress in his campaign against the autocratic rule of Maharaja Hari Singh. Soon after, with the support of the members of the National Conference, he adopted a manifesto entitled 'New Kashmir' which defined the political and economic objectives of the Conference vis-a-vis Jammu and Kashmir state. As a result of the manifesto, the popularity of the Sheikh touched new heights and he was addressed as the 'Sher-e-Kashmir' (The Lion of Kashmir) by the masses. Having achieved popularity, Sheikh Abdullah launched a campaign against the Maharaja in 1946 demanding that he should quit the state and hand over power to the people. The 'Quit Kashmir' campaign was more or less on the same line as the 'Quit India' campaign started by Mahatma Gandhi in 1942. The Maharaja acted firmly to crush the campaign. The Sheikh was arrested and sentenced to three years' imprisonment.

The National Conference, headed by Sheikh Abdullah had become a force to reckon with, and had by now eclipsed the Muslim Conference whose leadership was in the hands of Punjabi-speaking Muslim leaders like Chaudhary Ghulam Abbas and Sardar Mohammed Ibrahim. The Muslim Conference functioned in the state as if it were a branch of the All-India Muslim League and wanted the whole of Jammu and Kashmir to become a part of Pakistan. During the period of the 'Quit Kashmir' campaign, when Sheikh Abdullah and his followers were in prison, they got an opportunity to rapidly organize their party throughout the state. The rulers of the state also encouraged them to counter-balance the popularity of the National Conference. Thus, the revitalized Muslim Conference spread its tentacles in Jammu, Mirpur, Poonch and Muzaffarabad. In the Valley of Kashmir, too, it got sound support from the chief *mullahs* (clergymen).

Although Sheikh Abdullah had not consulted the leaders of the Indian National Congress or Mahatma Gandhi before starting the 'Quit Kashmir' campaign, yet Jawaharlal Nehru endorsed his stand. He stated that the Congress and the people of India were behind Sheikh Abdullah in his struggle against the autocratic rule of the Maharaja. To give moral support to the National Conference, he decided to enter Jammu and Kashmir state. The Maharaja did not permit him to do so and he was arrested at Kohalar Bridge near Rawalpindi, while forcing his entry into the state. Nehru remained in detention for a few hours and was released at the intervention of the Viceroy, Lord Wavell. Because of this support, the morale of the members of the National Conference was immensely boosted and Sheikh Abdullah shot into national and international fame.

The 'Quit Kashmir' campaign was started by Sheikh Abdullah in 1946 while the British Cabinet Mission was in Delhi to resolve the deadlock between the British government and the Indian people. The Viceroy, Lord Wavell wanted Nehru to be present in Delhi for the negotiations. As a result of these negotiations, an interim government was formed at the centre, headed by Nehru. On the resignation of Lord Wavell, Lord Louis Mountbatten was sworn in as Viceroy on

24th March, 1947. He was briefed by the Labour government of Britain to transfer power to Indian hands. On 3 June 1947, His Majesty's Government announced its intention to transfer power to Indian hands by June 1948 or earlier. With regard to the transfer of power to the Indian states, Lord Mountbatten in his address to the Chamber of Princes on 25th July 1947 stated that the Indian Independence Act would release the states from all their obligations to the Crown and technically and legally they would become independent. He warned, however, that if the states did not remain within either of the two dominions, chaos would result; they could not really function as independent entities.

Two state departments were set up, in India under Sardar Vallabhbhai Patel, with V.P. Menon as secretary; in Pakistan under Sardar Abdul Rab Nishtar with Mr Ikramullah as secretary. Although the states were theoretically free to link their future with either of the dominions, there were certain geographical compulsions which could not be ignored.

Of the 565 states, the vast majority were linked geographically to India. The Cabinet Mission Plan had proposed that the states should surrender to the government three subjects—defence, external affairs and communications.

On the advice given by Lord Mountbatten, almost all the states contiguous to India, signed the Instrument of Accession with the Dominion of India on or around 15 August, 1947, except Hyderabad, Jammu and Kashmir and two small states.

Jammu and Kashmir state refrained from signing, as Maharaja Hari Singh's relations with the Congress had never been cordial and he was advised against it by his counsellors. The state had common borders not only with India and Pakistan but also touched the borders of three countries—the USSR, Afghanistan and China. Again, while the ruler of the state was a Hindu, the majority of his subjects were Muslims. On the advice of his counsellors, he entered into a standstill agreement with Pakistan.

The districts of Sialkot, Gujrat, Rawalpindi and Jhelum of West Punjab have a common border with Jammu province of the state. Soon after the partition of 1947, the Hindus and Sikhs dwelling near the border areas crossed into Jammu for the safety of their lives. The Muslims of the border areas of Jammu did not migrate to West Punjab in the beginning, as conditions in the state were normal. Despite the fact that a large number of Hindus and Sikhs had crossed into the state from the border districts of West Punjab, Maharaja Hari Singh could not in the changing political scenario make up his mind about the accession of the state either to India or to Pakistan. Nevertheless, he released Sheikh Abdullah on 29th September 1947 in order to bring about a rapproachment with the National Conference which had become powerful in the Valley of Kashmir. Nearly two months had elapsed since the transfer of power and the indecision of the Maharaja regarding accession prompted Pakistan to arrange an invasion of the state with the help of tribal marksmen and retrenched army personnel, who, crossed over and pushed back the armed forces and police of the state. By the middle of October 1947, a large area of Mirpur and Poonch was under the possession of the tribesmen who, plundering and looting, reached within ten miles of Srinagar airport. The Maharaja had no alternative but to leave Srinagar and move down to Jammu. He sent messages to Lord Mountbatten and the defence ministry in New Delhi requesting military assistance. The Government of India could not help unless Kashmir acceded to India. The Maharaja thereupon, offered to sign the Instrument of Accession. The Government of India accepted the accession of the state temporarily, noting that the will of the people of Jammu and Kashmir would be ascertained for final accession at a future date, when the law and order situation had been restored and Pakistan forces had withdrawn.

The Indian army then entered Kashmir and pushed back the tribal raiders from Baramulla, Poonch, Rajauri and other sectors which had been occupied by them. When the Maharaja acceded to the Indian Union, and Indian troops were flown into Srinagar, Sheikh Abdullah and his National Conference organized an effective civilian resistance against the invaders from Pakistan. An interim government was set up and Sheikh Abdullah was invited to work in liaison with Mehr Chand Mahajan, the *Diwan* of the state. Soon Hari Singh abdicated in favour of his son Karan Singh, who was appointed Sadr-e-Riyasat, the constitutional head of the state. Sheikh Abdullah became the prime minister in 1948.

At the intervention of the United Nations Organisation, after a long and protracted wrangle, a cease-fire took place on 1st January 1949, between the Indian and Pakistani forces. As a result, 5,000 square miles, comprising Gilgit, North Ladakh, Ballistan, the small valleys of Mirpur and Muzaffarabad came under the control of Pakistan and this area has named 'Azad' Kashmir by them.

Soon after partition, a new political party, the Praja Parishad was formed in Jammu by the Hindus, many of whom were refugees from Pakistan. One of Sheikh Abdullah's first steps was 'The Big Landed Estate Abolition Act' which abolished *zamindari* limiting land-holding to twenty acres. As it happened, most of the big landlords in Jammu were Hindus who described this act of Sheikh's as communal. Many politicians in Delhi also shared this perception. Vigorous propaganda was launched by the Praja Parishad, duly supported by the Hindu refugees, against Sheikh Abdullah and his government for its policies which were dubbed anti-Hindu. In recruitment to the civil services, it was alleged, Hindus were being discriminated against; in matters of industrial and economic development, preference was being given to the residents of the Kashmir Valley which was predominantly Muslim.

The status and position of Jammu and Kashmir state in India was recognized by the Indian constitution to be different from that of all the other Indian states, which had been integrated into India. Under Article 370, the power of the Indian parliament to make laws for Jammu and Kashmir state were limited to:

(a) those matters in the Union list and Concurrent list which, in consultation with the government of the State, are declared by the President to correspond to matters specified in the Instrument of Accession governing the accession of the State and (b) such other matters in the said list as with the concurrence of the government of the State, the President may by order specify.

With India's complaint against Pakistani aggression pending before the Security Council, the state had become a subject of international controversy. India had also made the self-imposed commitment to a plebiscite, after the Pakistani army had been moved out of that part of the state territory which it had forcibly occupied. After the Cease Fire Agreement on January 1, 1950, efforts were made by the Security Council of the United Nations to find out ways and means of conducting a plebiscite in the two Dominions. Representatives were sent to discuss the matter with India and Pakistan in order to arrive at some mutual understanding but they did not succeed.

In the meantime, an announcement was made by the National Conference government that elections would be held in the state for the formation of the Constituent Assembly. The elections were to be held in September-October 1951. The National Conference put up candidates for all the seventy-five constituencies. The Praja Parishad candidates contested all the seats in Jammu and a few seats in the Kashmir Valley. On declaration of the election results, all the seats were found to have gone to the National Conference. The Praja Parishad accused the National Conference of having rigged the elections. Thereafter, it began to play a more aggressive role against the government of Sheikh Abdullah.

The Constituent Assembly of the state which also functioned as the Legislative Assembly was supposed to ratify the accession of the state to the Union of India but it did not do so for the next two years. In the meantime, the state adopted a separate flag of Jammu and Kashmir and resolved that the hereditary ruler could be the President, to be called *Sadr-e-Riyasat*, if he was duly elected by the legislature. There was some difference of opinion regarding the applicability of the fundamental rights contained in the Indian constitution to Kashmir. In July 1952, Nehru and Abdullah signed an agreement in Delhi in which it was specified that the state of Jammu and Kashmir, while part of the Indian Union, enjoyed certain unique privileges. The President of India could declare a state of emergency in Jammu and Kashmir only at the request of or with the concurrence of the state government. Citizens of the state had rights relating to land within the state which were denied to Indians from outside the state.

The political ideology of Abdullah had a distinct socialist tinge and one of his first priorities was land reform. By March 1953, he enforced a revolution in the land-holding pattern of the state and wanted to confiscate land without giving compensation to zamindars, whereas the Indian constitution guaranteed compensation. While Nehru was prepared to support the Sheikh, some of his colleagues and many opposition leaders were not.

One of the most vehement critics of Sheikh Abdullah and his policies was Dr Shyama Prasad Mookerjee who had resigned from Nehru's cabinet in 1950 to form the Bharatiya Jan Sangh, which stood for the protection of Hindu culture, traditions and interest and was mainly supported in north India by Hindu refugees from West Punjab, North West Frontier Province (NWFP) and Sind. The sufferings which they had undergone during partition made many of them anti-Muslim. Mookerjee established close links with the Praja Parishad and its leader Prem Nath Dogra and held that by having a separate head of state and a separate flag, Sheikh Abdullah was drifting towards the formation of an independent country. He, therefore, advised the members of his party to start an agitation against the Sheikh.

The movement caught the imagination of Hindus in Delhi, East Punjab and Jammu. Meanwhile, the Praja Parishad launched a movement for the integration of J & K state with India by proposing a common flag, a common head of state and a common Constitution. Hundreds of people were arrested for raising the slogan:

ek desh men do vidhan
ek desh men do nishan
ek desh men do pradhan
nahin chalenge, nahi chalenge

(Two constitutions, two flags, two presidents in the same country cannot be tolerated.) The Bharatiya Jan Sangh launched a protest movement in Delhi and East Punjab and particularly in the border town of Pathankot to impress upon the people of Jammu that the people of India were behind the Praja Parishad in their demands. Propaganda was unleashed in the press that Sheikh Abdullah was a turn-coat, he was pro-Pakistan, and was about to repudiate Kashmir's accession to India and that he wanted an independent Kashmir.

To thwart the movement of the Bharatiya Jan Sangh, Mridula organized a camp of the Congress workers of East Punjab in Pathankot which was the railway terminus of East Punjab and the only entry-point to Jammu and Kashmir state. The camp was attended by Prabodh Chandra, Sita Devi, Dr Prakash Kaur and many other prominent M.L.As and leaders of East Punjab. Some top leaders of the National Conference of Jammu and Kashmir state, such as Maulana Mohammed Sayeed Masoodi and Chaudhary Mohammed Shafi addressed the Congress workers and apprised them of the issues that faced the state. They explained the role played by the members of the National Conference in throwing back the tribal raiders prior to the advent of the Indian Air Force and in preserving communal harmony which finally led to the accession of Jammu and Kashmir state. On conclusion of the Congress workers' camp, Mridula impressed upon the leaders of the Congress the importance of organizing campaigns in the districts to which they belonged to counteract the propaganda of the Bharatiya Jan Sangh against Sheikh Abdullah and his government. Accordingly, in some towns of East Punjab, Congress Committees did organize parallel meetings. In order to enlist the support of all non-communal political parties, Mridula started a forum called 'Friends of New Kashmir Committee' on December 10, 1952 at Delhi with the assistance of Subhadra Joshi, M.P., Shah Nawaz Khan,

M.P., Amar Nath Chawla, Bishwa Bandhu Gupta, Sikandar Bakht, Shiv Charan Gupta, H.K.L. Bhagat, Om Prakash Bahel, Sakil Ahmed, Anees Kidwai, A.P. Dube, B.P.L. Bedi, M. Farooqi, Brij Mohan Toofan and others. A joint statement was issued which read as under:

The meeting unanimously condemned the Parishad agitation as anti-democratic, reactionary and communal, having as its aim the wiping out of the democratic gains of the common people of Jammu and Kashmir State, embodied in their new Constitution and accepted by the Indian Parliament. Further, in the opinion of the meeting, the aim of the Parishad agitation is to foist upon the common people of Jammu and Kashmir the old autocratic rule of the Maharaja and the *Jagirdars*, to deprive the peasant of his land and the people of their democratic rights. The Indian supporters of the Parishad agitation similarly represent the feudal and other reactionary interests, who are afraid of the democratic advance of the people. These reactionaries are trying to confuse the public mind in India and Jammu and Kashmir State by camouflaging their communal and politically disruptive intentions under the cry of full accession to India and such other slogans. The meeting considered it necessary to expose the reactionary communal character and aims of the Parishad agitation, to fight back the communal propaganda of their Indian supporters, to make known to the people of India the real facts of the situation and finally to popularise the democratic reforms embodied in the new Constitution of Kashmir.

A united front of all non-communal parties was thus formed to counteract the propaganda of the Praja Parishad and the Bharatiya Jan Sangh. Branch offices of the 'Friends of New Kashmir Committee' were set up at Pathankot, Amritsar and Jullundur.

For security reasons, entry permits were required to be obtained from the Government of India and the government of Jammu and Kashmir for entry into the state. Dr Shyama Prasad Mookerjee, the founder-president of the Bharatiya Jan Sangh took serious objection to this by asserting that there should not be any restriction for an Indian national to enter Jammu-Kashmir since the state was an integral part of India by virtue of its accession to the Indian Union in 1947. He emphatically declared that neither the Government of India nor that of Jammu and Kashmir state had any justification or valid authority to impose restrictions on the free movement of an Indian national within the territory of the Indian Union. To press his point, he announced that he would defy the orders of both the governments and enter Jammu and Kashmir without an entry permit. In early 1953, he set out for Jammu in the company of his colleague Vaidya Guru Dutt. There was a huge crowd to welcome him at Pathankot. They had the impression that he would be arrested by the Government of India within the territory of East Punjab. But that did not happen. He was freely allowed to move further and finally reached the river Ravi unhindered. After crossing the bridge, when he entered the territory of Jammu and Kashmir state near the Lakhanpur check-post, he was arrested by the state authorities along with Vaidya Guru Dutt and taken to Srinagar. This was done with the tacit approval of the Government of India as it wanted to keep the arrest of Dr Shyama Prasad Mookerjee beyond the jurisdiction of the Supreme Court of India as it was apprehended that the Supreme Court would have given a judgement in his favour and would have ordered his release. As the Supreme Court of India did not have jurisdiction over the territory of Jammu and Kashmir state, the arrest of Dr Shyama Prasad Mookerjee could not be challenged in that court.

In Srinagar, the prisoners were kept in a sub-jail at Nishat. Dr Mookerjee was not in the best of health. Moreover, the climate of Srinagar and prison-life and its surroundings did not suit him and his health began to deteriorate. As the state government had not issued any bulletin about the state of his health, the general public was not aware that he was not keeping well. After more than a month in prison, he suddenly collapsed on June 23, 1953. The news of his death spread like wild fire and a wave of resentment swept all over the country at the callousness shown by the government of Jammu and Kashmir state. All the opposition parties blamed Sheikh Abdullah, for suppressing news about his illness and for not extending proper medical aid to him. The Bharatiya Jan Sangh went to the extent of calling it a conspiracy to kill him.

To counteract the allegations of the Bharatiya Jan Sangh and the Hindu communal press, the 'Friends of New Kashmir Committee' organized public meetings to apprise the people that there was no conspiracy to kill Dr Mookerjee and that his death was due to natural circumstances. But their attempts to defend Sheikh Abdullah met with little success and there was a group even within the Congress which was not prepared to absolve Sheikh Abdullah for the negligence he had shown in saving the life of Dr Mookerjee. They criticized him both privately and publicly.

The widespread criticism of Sheikh Abdullah by Indian leaders and the Indian press made him extremely unhappy. He started believing that almost all Indians doubted his loyalty towards India.

A tense situation developed in India during July 1953 in respect of the Kashmir issue. Various rumours were being set afloat within Jammu and Kashmir state and within India. This has been aptly described by Mridula in a note addressed to Jawaharlal Nehru on 5 July, 1953:

In the context of Dr Mookerjee's death and the reported news of the differences in the ranks of the National Conference which have not been contradicted, doubts, it appears, have arisen in the minds of the common people about the future relationship of Kashmir with India. Lots of stories in this connection are being talked. It is said that Sheikh Abdullah is out to make Kashmir a sovereign independent State and is not prepared to implement the agreement he made in this connection.

It is further alleged that Bakshi Ghulam Mohammed on the other hand wants to implement the agreement and is for the accession of Kashmir to India. It is also a common talk that some days back about 50,000 Kashmiris took out a procession in Srinagar in which slogans, *'Pakistan Zindabad, Hindustan Murdabad'* etc. were raised. The procession, it is alleged, was organised by Ghulam Mohiuddin of the National Conference with the connivance of Sheikh Abdullah.

Again it is said that Chester Bowles and Adlai Stevenson met the Sheikh and encouraged him for an independent Kashmir. The story is further carried that when sometime back Pandit Nehru visited Kashmir, the National Flag was not hoisted and there was some trouble between Pandit Nehru and Sheikh Abdullah. The Sheikh, it is alleged, told Pandit Nehru that he was dealing with the Prime Minister of Jammu and Kashmir State and not a person like Dr Gopichand Bhargava.

It is again averred that when Maulana Azad visited Kashmir recently, the Kashmir audience, in a public meeting refused to listen to him and left the meeting after Sheikh Abdullah had concluded his speech and the Maulana was asked to speak.

Another story is that sometime back the officials of the Ministry of Communications went to Kashmir to take charge of certain things connected with communications, the Sheikh not only gave them no accommodation but, in fact, did not see them and that they have come back without doing anything.

It is being said that a very critical situation has developed there and the chances are that India would lose Kashmir. The general talk is, 'Kashmir is lost to India and the loss of Kashmir would shake the very foundation of the Congress Government in this country'. The talk goes to the extent that since Pandit Nehru has ben personally responsible for Kashmir Policy, the situation of the Kashmir Government is very much going to affect his future as leader of the Nation. The common man, therefore, seems to be quite confused about the Kashmir question at present and generally speaking the doubts seem to have arisen about Kashmir's future relationship with India.

Towards the end of July 1953, the political situation in the state became confusing. Abdullah in his speeches and statements, implied that accession to India had never been completed and the problem had to be solved to the satisfaction of India, Pakistan and Kashmir. This was the first time he brought in the necessity of Pakistan agreeing to any solution and his utterances amounted to saying that Kashmir should have an independent status. He also began airing all sorts of grievances against India, expressing doubts about India's secularism and accusing the government of being anti-Muslim. These speeches disturbed Nehru who wrote to Abdullah but the latter reiterated his viewpoint. Nehru then sent Maulana Azad to Srinagar in the hope that he would be able to persuade Abdullah to adopt a more reasonable attitude. According to Karan Singh , Abdullah ignored him and the National Conference workers insulted him.[1] Azad, it is said, returned convinced that Abdullah was not amenable to any logic.

In a speech delivered at Ganderbal on 2 August 1953, Abdullah said that Kashmir must steer clear between the two extreme views of merger with India and merger with Pakistan; there were dangers in both these stand-points and accession had to be ratified by the will of the people. Many of his colleagues including Girdhari Lal Dogra, Sham Lal Saraf and D.P. Dhar disagreed with him. Bakshi Ghulam Mohammed, the Deputy Prime Minister, spearheaded a movement against Sheikh Abdullah and reaffirmed his state's Accession to India as final. There was a sharp division within the rank and file of the National Conference which took a serious turn on 7th August. The Sheikh asked for the resignation of Saraf, Minister of Health, Development and Local Self-Government on charges of 'serious administrative complaints' regarding the working of his department and the latter refused. Bakshi, Dogra and Saraf sent a memorandum to Karan Singh accusing Abdullah and Beg of trying to rupture the relations of Kashmir with India.[2]

Thus, it became clear beyond doubt that sharp divisions had grown within the rank and file of the National Conference. These differences were not confined only to the rank and file members of the National Conference but were also apparent in the Working Committee of the National Conference, in the state cabinet as well as the Constituent Assembly. At a meeting of the Working Committee to discuss the implementation of the Delhi Agreement, Sheikh Abdullah was able to secure only three votes for his viewpoint against the fourteen cast for Bakshi Ghulam Mohammed, excluding Bakshi's own vote. Sheikh Abdullah strongly criticized Ghulam Mohammed Sadiq, President of the Constituent Assembly for the revolt against his leadership. Sadiq was told that he had no following in the National Conference which proved to be wrong, as a few days later, he was elected Chairman of the Reception Committee of the National Conference. It was at a cabinet meeting that Sheikh Abdullah had demanded the resignation of the Sham Lal Saraf.

Next day, Dr. Karan Singh concerned by these developments. invited Sheikh Abdullah to his residence. When Sheikh Abdullah narrated the incidents that had made him ask for Saraf's resignation, Karan Singh suggested that he and his Cabinet colleagues should come over and discuss the whole matter in depth. Abdullah, however, sidetracked the suggestion and launched into an angry tirade against the Indian press, saying that it was from this quarter that the most effective opposition to his rule was being organized. The meeting lasted for forty-five minutes after which Abdullah left for Gulmarg.[3]

On that night, a meeting was held at the residence of Karan Singh where Bakshi Ghulam Mohammed, D.P. Dhar, M.K. Kidwai, Chief Secretary of the J & K government, an ADC and a few pressmen were present. Karan Singh, D.P. Dhar and B.M. Kaul were all of the view that Abdullah should be dismissed. They felt that 'unless something was done to curb Sheikh Abdullah, the situation would steadily deteriorate and finally get completely out of hand with grave and incalculable consequences for the entire country. The Kashmir issue was still on the Security Council agenda and it would be disastrous if the Sheikh, who had twice been sent to the UN as a member of the Indian delegation, was to do a volte-face while still prime minister'.[4]

Convincing Nehru that Abdullah ought to be dismissed was not an easy task. According to General B.M. Kaul, it was D.P. Dhar who persuaded Rafi Ahmed Kidwai that the situation in Kashmir was worsening and Abdullah had to be removed. Kidwai wrote to Abdullah that he would like to visit Srinagar to discuss matters but the latter replied that no useful purpose would be served. Kidwai was naturally hurt by this and he and Dhar prevailed upon Nehru that strong action must be taken and the latter was finally persuaded that Abdullah wanted an independent Kashmir which was dangerous and hence must be removed but Nehru was insistent that Abdullah should not be arrested. Bakshi, however, made it clear that he could not run the government unless Abdullah and Beg were arrested.[5]

At 4.30 a.m. on 9 August, a notification was issued dismissing Sheikh Abdullah from prime ministership and dissolving his council of ministers. Nehru was complimented by Abdullah's opponents for his strong action and people felt that he would not have consented to the arrest of his close friend and associate unless there had been compelling reasons for doing so. There is a view that Nehru, in fact, did not want him to be arrested and this was one of the conditions he laid down before the dismissal and that the decision to arrest Abdullah was taken by Karan Singh, D.P. Dhar, Bakshi and Kaul without Nehru's consent since they knew that if Jawaharlal was told in advance, he would prohibit such a move. According to Kaul, when Nehru heard about the arrest, he rang up Karan Singh and blew him up. The latter heard only a part of the outburst, and, completely shaken, handed over the phone to A.P. Jain, who in turn, not being able to take the tirade by himself, passed the phone over to D.P. Dhar. Two days later when Kaul met Nehru in Delhi, Nehru vented his extreme annoyance that Abdullah had been arrested despite his orders to the contrary.[6] But S. Gopal in his book *Jawaharlal Nehru— A Biography* writes that Nehru was prepared for the dismissal of Abdullah on 9th August and quotes him:

For the last three months, I have seen this coming, creeping up as some kind of inevitable disaster. I did not of course know the exact shape it would take. To the last moment, I was not clear what exactly would happen.

B.N. Pande, in his book *Nehru* writes, 'when the final act took place in August 1953 and Nehru had to sanction Abdullah's deposition, his eyes filled with tears.' In this context, it would be worth quoting the observations made by Y.D. Gundevia in his book *The Testament of Sheikh Abdullah* which reads as under:

All the little people that want to claim credit for their great personal role in the arrest of Sheikh Abdullah in August 1953 have, surprisingly uniformly, waited patiently for the great man to leave us for good on that fateful day in May 1964, before they opened their mouths to tell us what Nehru said to them on this date and what he said about Sheikh being this, that and the other on that date and on and on they go. They waited long, till the ashes of the dead Prime Minister were scattered over India before they began telling their sorry tales.

Many, like Frank Moraes, felt that power had gone to Abdullah's head: 'Abdullah struck me as a highly egocentric individual, personally very ambitious'. When Moraes met him in 1951 Abdullah had been in power for three years, 'power had plainly infected his thinking and judgement... he talked disdainfully of New Delhi and insisted on the necessity of a special status for Kashmir. Since Kashmir already enjoyed a special status (Abdullah was the only chief minister to be designated prime minister) his conversation puzzled and disturbed me... In the hindsight of the after events, I have a feeling that even at that time his mind was moving towards independence for the Valley of Srinagar, with himself as the Kashmir equivalent of the Grand Mogul'.[7]

Sheikh Abdullah who was dismissed from prime ministership was arrested at Gulmarg on the morning of 9 August 1953 Mirza Mohammed Afzal Beg, his closest colleague was also arrested. The order of detention on Sheikh Abdullah was served on him at the famous holiday resort by a superintendent of police. He shouted at the officer who served him with the warrant of arrest. Begum Abdullah and her children were allowed to return to the official residence of the former prime minister at Srinagar. They had been permitted to stay there for some time. All possible precautions were taken for the protection of the family. Sheikh Abdullah was lodged in a beautiful palace, known as Tara Niwas in Udhampur. Mirza Afzal Beg was lodged in the Central Jail at Jammu.

Mridula had visited Srinagar many times before 9 August 1953 to acquaint herself with the latest political developments. She is said to have carried messages back and forth between Jawaharlal and Abdullah, trying to narrow down the difference between them. Sheikh sahib's arrest was a rude shock to her. On that day, she was in Amritsar in connection with her work regarding recovery of abducted women. On hearing the news, she rushed to Delhi by the first available train and on arrival, immediately got in touch with the members of parliament who represented J & K state. All of them owned allegiance to Sheikh Abdullah and were naturally upset by his arrest. She requested them not to discuss anything with the press until they had decided upon a policy among themselves in consultation with their leader in parliament, Maulana Masoodi. She also asked them to freely place their views before Nehru, and if need be before the Congress party, and try to make them understand the different viewpoints that prevailed in Kashmir. She cautioned then about rushing to conclusions without hearing the other points of view. She felt that there was a press campaign against Sheikh sahib, similar to that which was started in Pakistan after Badshah Khan's arrest in which he was completely misrepresented and misunderstood. The Pakistani media had accused Abdul Ghaffar Khan of working for an independent Pakhtoonistan and the government arrested him on charges of sedition. Mridula had championed his cause and urged that an impartial enquiry be held and in the meanwhile Abdul Ghaffar Khan should be released. Now she asked the Kashmiri MPs not only to take Maulana Masoodi's advice as National Conference workers but, also as members of the Congress party in parliament, and to consult Nehru. She wanted to play the role of a mediator between the Kashmiri members of parliament and Jawaharlal. According to B.N. Malik, Director of the Central Intelligence Bureau, Maulana Masoodi enlisted her assistance in forming a pro-Sheikh lobby in Delhi.[8] Discussions and consultations started among them to chalk out a programme for a future course of action. Since they were all staying in Constitution House, these meetings could be held frequently. Mridula firmly believed that for winning Kashmir over to India, the Government of India would have to turn to Sheikh Abdullah one day or another. In a statement issued on 13 August 1953, she said that to call Sheikh sahib a traitor, a foreign agent or a communalist was wrong and unworthy of those who stood for democracy.

On 10 August, 1953 Jawaharlal Nehru made a speech in the Lok Sabha:

Certain events have occurred in the State of Jammu and Kashmir with dramatic suddenness during the last two days and I am therefore, venturing to take some time of the House placing before it such facts as are known to us. Not only this House but the country at large must have viewed these developments with anxious concern. The State of Jammu and Kashmir has been to us not merely a piece of territory which acceded to India five and three quarters of years ago, but a symbol representing certain ideals and principles for which our national movement always stood and which have been enshrined in our Constitution. It was because of a community of these ideals and principles which brought the State, in a moment of crisis in October 1947, into the larger family of India. But even before that constitutional event took place, a devotion to these ideals and to certain common purposes had brought the national movement of Jammu and Kashmir State in tune with the struggle for freedom that inspired the people. In the Kashmir State, it was the National Conference which represented that struggle and spoke on behalf of the people there. The association of the State with India therefore had a deeper significance than even the constitutional link that was built up.

Much has happened during these years and we have faced trial and tribulation together. Even at the time of accession of the State to the Union of India, it was made clear that it was for the people of the State to determine their future when suitable opportunity for this arose. The union was a free union of free people without compulsion on either side. It was recognized from the very outset that the particular position of the State made it necessary for a special position to be accorded to it in our constitutional relationship. Later when the Constitution of the Republic of India was drawn up and finalised, this special position was recognized. It was made clear that any change in or addition to that position would depend upon the wishes of the people of the State as represented in their Constituent Assembly. The subjects of accession were three, namely, Foreign Affairs, Defence and Communications. In an agreement that was arrived at last year, known as Delhi Agreement, certain consequential and implied powers were defined. But the essential subjects of accession remained, the three already mentioned.

I mention this because much confusion has been caused by forgetting this basic fact that we have all along stood for a special position of the Kashmir State in the Indian Union. Some people have talked of a 'merger'. The word, of course, is totally inappropriate in any event and, to the extent it meant something beyond the constitutional position and our present agreements. Others advocated a weaker association which would have been against the basic policy that had all along been agreed to and would have involved grave difficulties.

In recent months, an unpleasant agitation sought to determine this basic position and created not only confusion but powerful reactions, more especially in the Valley of Kashmir. That has been one of the more important of the difficulties that the people of Kashmir and of India have to face. Unfortunately, some persons in the State were so affected by this agitation as to forget that community of ideals and principles which had brought Kashmir and India together. It was still more unfortunate that wrong advice was given by them to Sheikh Abdullah who had been the acknowledged leader of the National Movement in the State and the Prime Minister. Certain utterances of Sheikh Abdullah reflected this advice and created confusion in the minds of the people of the State. Disruptive elements, who had not accepted the principles on which the democratic movement in the State had been built up, took advantage of this position and attempted to disrupt the State. At a time when economic problems of grave import all over the State demanded attention and solution, the Government of the State was divided and ceased to function effectively.

A serious situation was thus created and there was a progressive tendency towards disruption. The Government of India were naturally gravely concerned at these developments but they did not wish to interfere, except with advice, in the internal structure and administration of the State. Advice was frequently given, but unfortunately it did not succeed in bringing about that unity which had shaken in the course of the past few months.

Some two weeks ago, two Ministers of the Kashmir Government, Bakshi Ghulam Mohammed and Mirza Afzal Beg, visited Delhi and had prolonged consultation with us. We pointed out to them the necessity for resolving their differences and working as a team in furtherance of the aims and objectives of the State. We assured them that we recognized the special status of the State and the Government of India did not wish to interfere in any way in internal problems which should be decided by the Government of the State. We were anxious to help, financially and otherwise, in the development of the State and were interested in the maintenance of security and internal order of the State.

A few days ago, we were informed that the differences within the Kashmir Cabinet had become even more pronounced and in fact ministers publicly spoke against and criticized each other and advocated rival policies. The majority in the Cabinet adhered to the objectives for which they had always stood. One member of the Cabinet, Mr Beg, however, progressively encouraged by Sheikh Abdullah, opposed these policies. A considerable majority of the Executive of the National Conference sided with the majority in the Cabinet and against the Prime Minister. The break was almost complete and it was impossible for the Government to carry on in this way. When we were informed of this and our advice was sought, we urged that some way should be found for the working of the Cabinet as a team on agreed principles and polices. This was an internal matter and we did not wish to interfere. Our interest was in a peaceful and progressive Government, having the support of the people. Three days ago, we learnt of the demand made by Sheikh Abdullah to one of his ministers to resign and the latter's refusal to do so. Events then moved rapidly. We were informed of them, but our advice was neither sought nor given. On the 7th August, the majority of the Cabinet presented a memorandum to Sheikh Abdullah in which they pointed out that factional tendency had been evident in the Cabinet and that this had been responsible for a progressive deterioration in the administration, that their advice had been disregarded and that the Cabinet, as constituted, could not continue. They communicated this memorandum to the Head of the State, the *Sadr-e-Riyasat.*

On the 8th August, the *Sadr-e-Riyasat* sent for Sheikh Abdullah and conveyed his deep concern at the serious differences which existed in his Cabinet. He impressed upon him the immediate necessity for restoring harmony and unity of purpose among the members of his Cabinet in the execution of his policy. Sheikh Abdullah could not give any assurance about the future and as to how he would be able to get over these differences. The *Sadr-e-Riyasat* therefore suggested that an urgent meeting of the Cabinet should be held at his residence that evening so that the possibilities of securing a stable, unified and efficient government for the State could be jointly explored. Sheikh Abdullah, however, did not agree to this. Later in the day, the *Sadr-e-Riyasat* wrote to Sheikh Abdullah pointing out all these facts and stating that in these conditions he had been forced to the conclusion that the present Cabinet cannot continue in office any longer and hence he had decided to dissolve the Council of Ministers. A formal order to this effect was passed and a copy of it was sent to Sheikh Abdullah.

Immediately afterwards, the *Sadr-e-Riyasat* called upon Bakshi Ghulam Mohammed to form a new Cabinet. In doing so, he stated that the continuance in office of the new Cabinet would necessarily depend upon its securing a vote of confidence from the Legislative Assembly during its coming session. Bakshi Ghulam Mohammed accepted this invitation and was sworn in as Prime Minister of the State.

I received information of some of those developments at 11 p.m. on Sunday night, that is night before last. Further information followed on Sunday morning.

Sheikh Abdullah had meanwhile gone to Gulmarg. In fact, the last order was served upon him in Gulmarg. Later he was placed under detention and so was Mr Beg. I have not received the exact papers in regard to this matter yet, but I understand that this was done in the interest of the peace of the State which was threatened in various ways.

Some time after the news of Sheikh Abdullah's arrest became known yesterday morning, small processions in protest started in some parts of Srinagar and converged towards Amira Kadal which is a bridge. Those processions became violent in some places and threw stones at the police and militia. On two occasions, the police had to fire on the crowd, it is stated, in self-defence. Three rounds were fired on one occasion and four rounds on another. The total casualties were three killed and one injured. The dead body of one person was paraded through the streets.

As it was Sunday, shops were generally closed, and there was little obstruction to traffic. There was no communal incident of any kind. So far as is known, there has been no trouble in any of the outlying areas. By the evening, the situation had improved considerably. Till last night, 35 arrests have been made. The Indian Army personnel was not involved in any way. The situation was dealt with by the Jammu and Kashmir police and militia. One party, however, of the Central Reserve Police functioned in one place.

Sheikh Abdullah was taken to Udhampur where he is lodged in the Rest House and every comfort has been provided for him.

It is a matter of deep regret to me that Sheikh Abdullah, an old comrade of 20 years, should have come in conflict with our other comrades in Kashmir and that it should have been considered necessary by the Kashmir Government to place him in detention for time being. I earnestly trust that this is a passing phase and that the leaders of Kashmir will cooperate together in the service of that beautiful and unfortunate land.

Last night, Bakshi Ghulam Mohammed, the new Prime Minister, broadcast a long speech in which he has referred to the recent developments as well as to the policies which he and his Government intend pursuing. I should like to repeat that we have considered these recent developments in Kashmir as an internal matter with which we should interfere as less as possible. On the large issues our policy remains what it was and we shall stand by the assurances we have given.

To the Members of this House, to the press and this country and the people generally, I would like to make an earnest appeal to exercise forbearance and restraint in regard to those events which have followed each other in quick succession in the Jammu and Kashmir State. We must send our full sympathy to the young *Sadr-e-Riyasat* to the Government and the people of that state who are facing this crisis, and assure them of all help that we can give them to bring about normality and a progressive administration which will serve the cause of the people of that State.

Championing of the Sheikh's Cause

The members of parliament representing Jammu and Kashmir state were extremely perplexed by the Prime Minister's speech and were in a great fix as to what stand should be taken by them. At that crucial stage when they contacted Mridula to seek her advice, she suggested that they should seek advice from their leader, Maulana Mohammed Sayeed Masoodi who was also the general secretary of the National Conference. She added that they should also talk over the matter with Jawaharlal Nehru who was the leader of the Congress Parliamentary Party in the Lok Sabha and as such was their leader also. Maulana Masoodi agreed to her suggestion and also agreed to express his views before the Congress Parliamentary Party. While addressing the party, Maulana reiterated their loyalty to India but expressed anxiety as to how the present link of accession between India and Kashmir could be maintained. He emphatically said that all the leaders of the National Conference, whether belonging to the Bakshi group or the Abdullah group, would continue to work as they had done before, for the maintenance of permanent India-Kashmir ties of friendship and association. He added that the internal changes in Kashmir were their own domestic affair and that a policy of outside pressure or interference at that juncture would end in disastrous consequences. He added that the recent changes in Srinagar did not 'affect a bit' the position of the Sheikh. 'He was leader yesterday, he is leader today and he will be leader tomorrow, and for winning over Kashmir to India, we will still have to turn to Sheikh Abdullah one day.' Mridula issued the following statement to the Press on 13 August, 1953, just four days after the arrest of Sheikh Abdullah:

Much has been said about the recent tragic events in Kashmir State, but what hurts one most is the manner in which steps taken for alleged ideological differences are being justified by making serious allegations against those who today are detained and cannot defend themselves. It is most unfortunate that through rumour-mongering and one sided publications, the national standards are being allowed to deteriorate to the same standard as in some other countries, where there is lack of toleration towards political opponents. Let us not forget that the tragic events in Kashmir are not the doings of a few weeks. They are the result of a series of planned events and it is only the historians of the future who will record who was at fault and responsible for them.

Today one is reminded of the torrent of campaign that the Ministry of North West Frontier Province in Pakistan started against that brave leader, Badshah Khan (Abdul Ghaffar Khan). They tried to hide their power politics on the ground of 'national security' and dubbed him as 'an Indian Agent'. Let us not degrade ourselves and follow such a path and swallow and believe all that is being said about Sheikh Sahib (Sheikh Abdullah). To depict him as a traitor, as a foreign agent, as a betrayer of the cause he preached and stood for, as a communalist is wrong and unworthy of those who have resented such treatment in other countries. It would also amount to undermining the effort of the nation to build up democracy on the basis of tolerance and understanding. The manner in which the communalists and reactionary forces have reacted is sufficient to show that there is something much deeper than is known to the public. I would, therefore, plead with the leaders, colleagues, the press and the people to rise to the occasion and behave as decent opponents if they so desired and, if charges are to be levelled, let it be done at a Court of Justice and not behind the back of those who cannot defend themselves.

No one in India apart from Mridula had the courage to denounce the actions of *Sadr-e-Riyasat* and Bakshi Ghulam Mohammed, the newly appointed prime minister of Jammu and Kashmir State. Besides, most thought that Nehru would have never given his approval for the arrest of Sheikh Abdullah unless there were valid reasons for it. All sorts of stories portraying the betrayal by Sheikh Abdullah began to gain currency. The most damaging was the one that Sheikh Abdullah along with the members of his family would have escaped to Pakistan but for the timely action of *Sadr-e-Riyasat* and Bakshi Ghulam Mohammed. In these circumstances, the issuing of such a statement by Mridula in support of Sheikh Abdullah invited the wrath of the general public and the governments of India and Jammu and Kashmir. Within hours, her reputation was jeopardized. All kinds of accusations were hurled at her and she fell from grace even in the Congress circle. A good number of M.P.s of the Congress Party brought pressure on Jawaharlal Nehru that she should be asked to vacate her rooms in Constitution House which was owned by the government. The aim and objective of these legislators was to isolate her from public life and prevent her from associating with the members of parliament who represented Jammu and Kashmir. Nehru acceded to the request of the Congress members of Parliament. As the central office of the

Recovery of Abducted Women was also functioning from Constitution House, she eventually disassociated herself from this humanitarian work. She also distanced herself from the Shanti Dal and the United Council for Relief and Welfare (UCRW).

On being relieved from these organizations, Mridula devoted herself completely to Kashmir affairs. Having been in the freedom movement since 1930, she was fully aware of the hardship and agonies political detenues' underwent. The hardest blow would fall on the detenues' and families as they were isolated not only from their relatives, friends, and well-wishers but also from their breadwinners. So, as a first step, she called on Begum Abdullah and assured her of all possible assistance. This boosted the Begum's confidence and she began to feel that she had found a real benefactor who was also very close to Jawaharlal Nehru. Apart from establishing close contacts with Begum Abdullah, she was looking for an opportunity to get in touch with Sheikh Abdullah. That opportunity came on the day of the Diwali festival when she conveyed her greetings to him in writing. Quick came the reply from Sheikh Abdullah in his letter dated 7 October 1953 which reads as under:

I thank you for your Diwali greetings which I heartily reciprocate. We are all well. My family came to see me and they stayed with me for a couple of days. Please convey my good wishes to friends if they are still left. Remember me to your family members.

The sentence, 'Please convey my good wishes to friends if they are still left' was meant for Jawaharlal Nehru. It also reflected his bitterness and sense of isolation. Mridula immediately wrote to him informing him that she was arranging to send him *Shankar's Weekly* the *National Herald* and the *Tribune*. She also enquired whether he would like any good books or some scientific and educational toys such as a mechano set. She had experienced jail-life and knew how depressed and lonely one could feel.[9] She was in touch with Begum Abdullah and requested her to send the children to her for their winter vacation.

Sheikh Abdullah replied that he would very much like some fiction and also some good historical novels. His companions could not read English, so he wondered whether she could send some Urdu novels and short stories by Prem Chand, Kishen Chander and Ismat Chugtai. He said that she should not talk of her prison days.

Those were the days of India's slavery. Now we are free and one should cheerfully bear all the hardships that may come in the way and should not crave even for ordinary human facilities. In this age of democracy, who would not like to while away his time in some useful activity but I wonder if those in authority will permit it. All the same, I thank you from the core of my heart for your thoughts about me, I was particularly touched by your reference about my children and wife...[10]

Sheikh Abdullah had never expected to be treated in this manner by Nehru and was deeply hurt that all his friends in India including Jawaharlal, had turned against him. He was, therefore, specially grateful to Mridula for the love and affection that she had been showering upon them in their distress.

My only regret is that it is not possible for me to establish any contact with you or for that matter with anybody inside or outside the State except my family members. The authorities have completely isolated us from our friends and well-wishers.

'The Tribune' and the 'National Herald' which you had so kindly arranged to be sent to us are also being withheld. Under these circumstances you will, no doubt, understand my difficulties in conveying to you and to my friends in Delhi my sincere thanks and affection for all their efforts... in clarifying our position and placing the true facts before the people of India.

He firmly believed that once the real facts came to light, his opponents would be doomed for ever.

You all known the character, aim and ambitions of the two main actors in this whole tragic drama-Bakshi and Sadiq. Nothing is hidden about them specially from Maulana Sayeed. He knows them in and out. Our friends despite their best efforts have not so far succeeded in silencing you all because they are powerless... I hope and trust that you will all continue your efforts in enlightening those who count in Delhi about the real facts of this tragic drama.... If Bakshi and Sadiq could not prove faithful to all their erstwhile colleagues of twenty years standing and more so to me, who was instrumental in raising them to this status, will anybody in his full senses ever believe that these two gentlemen will remain loyal to Panditji and to India?... They will never succeed in creating the union of souls which Panditji often declared was his objective... The Muslims of Kashmir expected not only justice and fairplay from Bapu's India, but even generosity. But, instead they got bullets.

No proof had so far been furnished as to why they were in prison and of the conspiracy that they had supposedly entered into with a foreign power for creating an 'Independent State'...

The means that are now being adopted in making Kashmir a citadel for secular India may well console the spirit of Stalin but never that of Mahatma Gandhi.

He hoped that she would continue to expose the false colours that these traitors of Kashmir had adopted in order to gain their unworthy aims.[11]

Mridula established close contact with almost all the members of Sheikh Abdullah's family. Apart from extending moral and financial support to them, she took upon herself all responsibilities which she could possibly shoulder. She looked into their day-to-day problems and tried to solve them as best as she could. She paid special attention to the education of the children as she did not want their education to suffer because of the detention of their father in jail. At that time both the sons of Sheikh Abdullah, Farooq and Tariq were studying in intermediate classes. Farooq wanted to study medicine in some medical college in India. Mridula got him admitted to the Medical College in Jaipur at Rajasthan. The Chief Minister of Rajasthan, Jai Narain Vyas, held Sheikh Abdullah in high esteem. Prior to the attainment of freedom, Nehru had made Sheikh Abdullah the President of the All India States' Peoples Conference and Jai Narain Vyas its General Secretary. Since then, a close friendship had grown between the two leaders. Mridula, therefore, thought that Jaipur would be the ideal city for Farooq Abdullah to pursue his studies. This proved to be correct. On a few occasions, Farooq felt uncomfortable in the Medical College of Jaipur where he was ragged by some doctors and students who subscribed to the ideology of the Rashtriya Swayamsevak Sangh (RSS). On the whole, however, his stay in Jaipur was quite congenial. At the request of Begum Abdullah, she also acted as a guardian to Farooq and Tariq when they were outside Kashmir, either in India or abroad. In fact, her house in New Delhi on Rajdoot Marg, Chanakyapuri, where she had moved after vacating Constitution House, became their transit home. On a few occasions with her assistance, they called on Prime Minister Nehru and even stayed at Teen Murti House as his guests. Sheikh Abdullah writes in his autobiography that whatever attitude Jawaharlal adopted on a political plane, at the personal level, he behaved like a gentleman and took a keen interest in the welfare and education of his children. In 1955, in a speech in the Lok Sabha, Nehru said that he still had personal regard and affection for Abdullah for which the latter sent him a letter of thanks. To no small extent, Mridula was responsible for the continuing good relations between Nehru and the Sheikh family.

The bunch of letters exchanged between Mridula and Farooq Abdullah during this period reveal the keen interest taken by her in his education and in solving his problems. She contributed in a big way to making him a doctor. In a letter to her written from Medical College, Jaipur, Farooq expressed his wish to get into the All India Medical Institute in Delhi so that he could be near Panditji:

You know very well, *Bahenji,* that Panditji is the only hope we have in India, after all the troubles I have gone through in the Medical College, Jaipur, because of my being Sheikh Abdullah's son, are so great that it makes me mad at times, so much so that I may have to leave my studies and this country. I cannot stand any more of these troubles merely because people believe that my father is a traitor. Time alone will show who is a traitor... I am anxious and worried. Kindly help me out.[12]

Mridula wrote to Dr. Zakir Hussain, Governor of Bihar to look after Farooq when he visited Patna and to V.V. Giri, then Governor of U.P. to look after him in Lucknow.

After finishing his MBBS, Farooq expressed his desire to do postgraduate studies in medicine in England. It was Mridula who wrote to Dr Jivraj Mehta, India's High Commissioner to the United Kingdom, about getting Farooq admitted to some hospital in the U.K. She also made arrangements for his funds for two months after which he was to be attached to a hospital there. From London, he wrote to her:

I must admit India House has been extremely kind to me and my family. We will never forget the sacrifice you are making for the poor people of Kashmir and the truth.[13]

In another letter he wrote:

You have been a source of great help to us. It reminds me of that great battlefield of Mahabharat in which there were great armies on one side and on the other side, there were only five people (Pandavas) and they were helped by Lord Krishna... You have been to us what Lord Krishna was to the Pandavas.[14]

Mridula's letters to Begum Abdullah, Farooq, Tariq and others reveal how deeply she had involved herself in the affairs of this family. When Begum Abdullah's mother passed away in Delhi and she could not come because of snowfall and disruption of traffic between Srinagar and Delhi, Mridula rushed to Maiden's Hotel at midnight and helped in the funeral plans. She kept all members of the family especially Begum Abdullah informed of the developments and the arrangements made for the funeral through letters and the telephone. Mridula kept in close touch with Begum Abdullah. The latter expressed her gratitude in one of the many letters she wrote:

Believe me, I feel short of words when I write to you, to thank you from the bottom of my heart for all the love and affection and hospitality you have always shown towards my children and myself. We feel we have a home in India, a real true friend in India and that is your home and you. Amen grant you a long happy life and victory over evil and injustice along with us all who are in the same boat with you.[15]

Even today she is full of gratitude for what *Bahenji* did for them:

She was like my sister. She never thought of herself but only of what she could do for us and others. She was utterly selfless.[16]

Mridula also helped the families of political leaders such as Mirza Mohammad Afzal Beg, Soofi Mohammed Akbar, Khwaja G.M. Chikan, Khwaja Mubarak Shah and others who were in jail. There was hardly any family of political sufferers in the Valley who did not benefit by her generosity, and innumerable people with some problem or the other were helped by her. Begum Abdullah said that she had 'a heart of gold' and was ready to help anyone in distress.[17] Her house in Delhi, said Dr Farooq Abdullah, was like a 'Kashmir Sarai'.[18] She ordered the most expensive food for her guests, provided them with transport and they used her telephone and other facilities freely. G.M. Shah writes:

Behenji's house became the centre of political activity of Kashmir. They wanted that justice should be done to Sheikh and thousands of others who were languishing in jail for no fault of theirs. She, as a noble freedom fighter, kept her house at the disposal of us all and provided food, lodging, medical aid, transport to all those who stayed at her house. I have seen myself that even the servants got the same treatment at her house as other members of her family. I went to Delhi after my release from detention in 1960 to meet Jayaprakash Narayanji about the Kashmir Conspiracy Case. I was unwilling to stay at her house, because I am a fanatic non-vegetarian by habit and by taste. I told this fact to her. She at once asked her Manager to have a non-vegetarian kitchen in her house. That demonstrated her magnanimity and concern she had for us all.[19]

Sadur-Ud-Din Mujahid, president of the Kashmir Freedom Fighters' Association and Chairman of the J & K Khadi and Village Industries Board, writes that when he was released from jail on parole on account of his illness, Sheikh Abdullah advised him to go to Delhi for a medical check up and stay with Mridula.

Next day when I reached Delhi Railway Station, along with my brother-in-law and son, Jenab Peer Maqbool Shah Geelani, Sajadaa Nisheen Khanyar was there to receive me. He caught hold of my hand and led me a few step towards a lady in *khadi*, who took my hand and told me that now I need not worry and everything will be alright. This was the first time I met Mridula. She took me in her car to her residence, where I saw Shri Kashyap Bandhuji, Hakim Ghulam Mohi-ud-Din and Shri Abdul Jabbar Tailor Master and some other supporters of *Sher-e-Kashmir.*

I stayed in the house of Mridulaji, was looked after by her to the extent that even minor details regarding my health were taken care of. She looked after my companions, consulted doctors, bore all our medical and other expenses. The love and affection she bestowed upon me was beyond my expectation. Only a loving sister would take such pains. She never, for even a moment, let me feel that I was not in my own home. She did not allow me to return to Srinagar till I had fully recuperated. Though I wanted to get my son Shakeel operated for tonsils, he showed reluctance and Mridulaji advised me not to press. Before we left for Srinagar from Delhi, *Behenji*, as she was generally called by all Kashmiris, gave me more than two dozen books, two khadi suits for me and my son, some medicines and mangoes. She bought First Class return railway tickets for me and my companions. She herself came to see me off at the Delhi Railway Station.

During my stay at her house, I observed that she had great love for humanity and tried to help each and every person in any way she could. Besides myself, there were a number of people whose health had deteriorated due to prolonged imprisonment. She always took them home, nursed them and provided them with the best treatment available, bearing all the expenses. She had made it a point to help every Kashmiri businessman or the like, who may have some work in any office or institution at Delhi. She was a true Gandhian in her personal life and followed the teachings of Gandhiji in her day to day life.

Regarding her house I can say only this much that this house had become a refuge to Kashmiri people, where there was some one who realised their plight and wanted to help them in their hour of need. There was no bar in entering her house because she was ever ready to come to the rescue of any one who came to her. I, for one, found her house the only place where one would get peace of mind and she would never let one feel that it was her house. She got real joy in giving and not taking and in her heart, depressed and socially backward people had a special place and this one could feel as soon as one would enter her house, which was always haunted by CID personnel.[20]

Chaudhary Muhammad Shafee continued to stay in her house in New Delhi while she was in detention in the Retreat in Ahmedabad. From there she sent a telegram to Mubarak Shah who had just been released from jail.

Greatly relieved, delighted, all facilities at Delhi at your disposal. Suggest your proceeding earliest for medical check up treatment at Medical Institute.[21]

Maulana Masoodi was acquitted from the Hazratbal case on medical grounds as he was suffering from T.B. He was advised to spend the winter out of Kashmir. Mridula invited him to Delhi and rented a house for him.

Meanwhile, on the political front, the detenues who were in prison wrote a letter to the chief secretary of Jammu and Kashmir State:

It is now nearly a year that we have been kept in detention in the various jails of the State. During this long period, we have not been apprised of the reason or reasons which could have formed the pretext for permitting such long incarceration of the persons who are convinced of their innocence. No opportunity so far has been offered to us to answer any allegation which could be made up against us or to explain our position with regard to our continued detention. If the authorities really believe that they have any allegations against us, it would have been but fair to bring us to a trial in a court of law. The act under which our detention was ordered, in the first instance, limited the period of detention to a term of two months at a time, creating a presumption that at the end of each period of detention and at the time of sanctioning a further extension, the authorities would have an opportunity of reviewing the case of each detenue. It is time that the authorities be moved to allot funds for payment to us of suitable family allowance for the maintenance of our dependents to save them from unmerited hardship.

A copy of the petition was smuggled out to Mridula who sent it to the Indian press on 5 September 1954 but the newspapers did not publish it.

The Standing Committee of the All India Newspapers Editors' Conference was scheduled to meet in Srinagar in September, 1954. Fearing that Bakshi Ghulam Mohammed would give them a false account of what was happening in his State, Mridula issued a communication to them giving them what she considered was the true state of affairs. According to her, the coup of 9 August 1953 was the result of a well-planned conspiracy by some power seekers. The atmosphere before the coup was deliberately created by them and the Indian press had been fed with false stories. She appealed to her journalist friends to find out the truth, and investigate to what extent the Bakshi government was as popular and democratic as it claimed to be.

Under Mridula's instructions, a deputation of the members of the National Conference who were followers of Sheikh Abdullah submitted an advance copy of a memorandum to the members of the Standing Committee of the AINEC (All India Newspapers' Editors' Conference). It was written in the name of 'The Oppressed People of Kashmir' and complained of their loss of civil liberties and the injustice done to their leader Sheikh Abdullah and the treachery of Bakshi, Saraf and Dogra. The memorandum described in detail how Bakshi, Saraf and D.P. Dhar had hatched a plot to oust Sheikh Abdullah and held that the Indian press had received either a garbled version or concocted facts on the Kashmir situation. The memorandum stated:

By 10 o'clock on the 9th August (1953), the Valley of Kashmir was occupied by the military, police, CRP and militia. Every street, road and lane, every town and village was under their charge. As Sheikh Abdullah was being whisked away with a heavy guarded military escort to his future prison, the radio had broken this tragic news and stunned the people. The whole Valley was in mourning, reeling under severe blow, mammoth processions emerged from villages and towns, raising the ear-splitting cries of 'Sher-e-Kashmir Zindabad'. A reign of terror was let loose. The streets were littered with dead bodies. The hospitals were filled with injured victims of promiscuous killings, though only a few could be carried to government hospitals for obvious reasons. Unfortunately, no complete record of the casualties admitted in the hospitals was allowed to be maintained. Even those who were dragged to prisons were a woeful sight. The tragedy became acute when the press, local as well as in India, preferred to observe complete silence about this wild beastliness. This is the tale of suffering that we have related to you. We are confident that you will extend to us your cooperation and support our demand that:

(a) The false accusation made against Sheikh Saheb be cleared and the people of India informed of the real facts and the release of all those who are detained without trial;

(b) The people of Kashmir demand an enquiry into the atrocities committed on them during the last one year;

(c) The suppression of liberties and the freedom of press and of expression practised in the state today is opposed vigorously;

(d) The people demand that the present regime which is nothing but the worst form of Police State be put an end to.

The objective of this was to challenge the myth of Bakshi Ghulam Mohammed that peace and tranquility reigned supreme in the Valley of Kashmir despite the detention of Sheikh Mohammed Abdullah in jail. In fact, it was at the instance of the government of Jammu and Kashmir that the Standing Committee of the All India Newspapers Editors' Conference had agreed to hold its meetings in Srinagar. the invitation had been extended in the hope that Bakshi Ghulam Mohammed and his government would get wide publicity in the Valley as well as in other parts of India.

Mridula and her co-workers wanted to present a grim picture of the Valley which, according to them, was groaning under the repressive policies adopted by Bakshi Ghulam Mohammed and his government. It is difficult to say what impact it created on the minds of the members of the A.I.N.E.C. since the Indian press had supported and praised Bakshi Ghulam Mohammed for saving the situation in Kashmir and their meeting in Srinagar was in continuation of that support to him.

For about a year after Sheikh's arrest, Mridula did not associate with active politics in the State. She had hoped that a rapprochment would be arrived at which would lead to his release; this proved false. On the contrary, the government of Jammu and Kashmir, headed by Bakshi Ghulam Mohammed, started a virulent propaganda against Sheikh Abdullah, dubbing him a traitor and widely publishing a story that he had a secret meeting with an American agent who had assured him full support in the fulfilment of his dream of 'Independent Kashmir'. Adlai Stevenson had met Abdullah and this was perhaps the basis of this rumour. To give currency to the story, a pamphlet entitled, 'Conspiracy in Kashmir' in the name of the 'Social and Political Group' was distributed in Kashmir and also to members of Parliament, office bearers of the Congress and many others. The summary of the pamphlet also appeared in some Indian newspapers. Mridula did not believe in the bonafides of that document and discovered that Ghulam Mohammed Mir Rajpouri and Manohar Kaul were the co-authors of that pamphlet and were known in Kashmir as opportunists. This pamphlet was later published in a booklet form, presumably by the Bakshi government, and widely circulated. Mridula promptly issued a rejoinder entitled 'Whose Conspiracy?' in which she rebutted the charges against Sheikh Abdullah and argued that not only was Abdullah being personally wronged but the way it was being done was creating bad precedents and endangering the political life of India. She strongly denied that there was any conspiracy between Sheikh Abdullah and any American agents.

She sent copies of the rejoinder to all members of Parliament, office-bearers of political parties, editors of daily and weekly newspapers and to various other eminent persons. Sheikh Abdullah was, naturally, very grateful and wrote:

I am extremely thankful to you that you have taken it upon yourself to fill up the gap. The way my life-long comrades have treated me would have left me completely disillusioned about the essential and basic values of humanity had not you proved a shining star in all this darkness. I am sure that Bapu's soul must be feeling proud of you. Please accept my heartfelt thanks for the affectionate way you and yours are treating my family and for the boldness with which you are holding aloft, in spite of numerous difficulties, the Torch of Truth. I have no doubt in my mind that the Truth will triumph and the dawn will follow the darkness. Our friend J. (Jawaharlal) will realise that he has been a victim of the foulest conspiracy to bring him down from the highest pinnacle of Truth to which Gandhiji had lifted him.[22]

Within a short time after the arrest and detention of Sheikh Abdullah, the law and order situation was brought under control by the Kashmir government with the assistance of the Centre. As the situation improved in the Valley, the flow of tourists also increased. Satisfied with the progress, the state government permitted journalists, politicians, officials and others to visit the state freely and see the situation for themselves. The favourable reports from the journalists evoked a sharp reaction from Mridula in the form of a circular letter entitled 'Visitors to Kashmir in Search of Truth' issued on 1st September 1954. In this she said that tourists and visitors were being given a false impression and were victims of mischievous propaganda. On 16 September 1954, she issued yet another circular letter in which she held that agent provocateurs carried on pro-Pakistan activities coupled with shouting the slogan'Sher-e-Kashmir-Zindabad' thus misleading the local population.

Most Indians held Sheikh Abdullah responsible for the Kashmir situation but Mridula said that they did not know what had been going on behind the scene, that they were hearing only one side of the story. Sheikh Abdullah's version had not been made known to the public and she considered it her duty to do this. The detention of Abdullah in the Tara Niwas Palace jail continued beyond 1954 and as the years passed by, Mridula stepped up her efforts to get him released. The more Bakshi tried to justify the continuation of his detention, the more she refuted the charges and missed no opportunity of denouncing his actions.

During one of his visits to Jullundur, Bakshi, reportedly said that 'Kashmir was based on Bapu's principles of truth, justice and humanity'. Mridula was furious and asked him not to compare his methods with those of Gandhiji. She requested Nehru to visit Kashmir and see things for himself and if he could not go, to send a renowned Gandhian like Vinoba Bhave or Satish Chandra Das Gupta or U.N. Dhebar, president-elect of the Congress, to see whether the ways and means adopted by Bakshi and his regime were in any sense Gandhian. In her opinion the government was in fact violating every democratic norm and heading towards fascism.

Mridula urged that either Abdullah be released or tried because his detention was based on falsehood and violated the Indian Constitution. The so-called conspiracy should be unearthed by a judicial enquiry. Mridula wanted the 'Preventive Detention Act' which empowered the government to detain a person for five years without a trial or without being charge-sheeted to be abolished. She accused the 'leader' of the Kashmir Legislative Assembly of having secured a 'vote of confidence' from the Assembly as well as the National Conference through fraudulent means. She pleaded that the vote of confidence should be taken in the presence of Jawaharlal Nehru or let the assembly be dissolved and fresh elections be held. She accused the National Conference of having bogus members on its rolls and of censoring news and communication with India.

Bakshi Ghulam Mohammed formed an organization called the 'Peace Brigade' whose task supposedly was to help the government in the maintenance of peace and order. Mridula accused it of being 'an organisation of hirelings, a number of whose members were previous convicts and known criminals', whose main job was to crush ruthlessly all opposition to the Bakshi government. Its leader, Ghulam Qadir Gandharbali was according to her 'the most notorious and brutal officer of the police force of J & K State'.

Mridula went to Anantnag on 25 May 1955 to investigate an incident in which people had been beaten up and women molested because they had accorded a warm welcome to Mirza Ghulam Mohammed Beg who had been released from detention the previous day. What happened at Anantnag was repeated at Sopore where members of the Peace Brigade harassed Praja Socialist Party members. She said that curbing of civil liberties on the one hand and hool-

liganism of the Peace Brigade, on the other, were producing hatred
for the Indian government, which was believed to be conniving with
all this. For the prestige of India to be restored in the minds of the
Kashmiris, restoration of civil liberties was essential.

In order to keep Indians better informed about Kashmir, Mridula
circulated a report in which she contradicted the impression that
peace and tranquility prevailed in the Valley. According to her:

a reign of terror has continued unabated... Persecution is going on with un-
mitigated fervour, unrelieved by any avenue of freedom... undertrials are being
kept in separate cells, that is, in solitary confinement... what pains most is the
utter affront to human considerations. If political vendetta is to become the
criterion, it will move in a vicious circle unceasingly... both Bakshiji and his
colleagues are suffering from an obsession which seems to weigh heavily
upon them. The obsession is Sheikh Abdullah.[23]

She had a long and frank talk with Dr Karan Singh whom she
apprised of the activities of the Peace Brigade and the National Conference. According to her, he was amazed to learn about some of the
events which she narrated to him. She tried to impress upon him that
Abdullah's followers were not at all anti-Indian or pro-Pakistan, that
the Sheikh if released, would not lead his followers towards Pakistan
and that the Bakshi government was deliberately distorting facts. The
Sadr-e-Riyasat, however, made it clear that the accession of Kashmir
to India was an accomplished fact and that the talk of a plebiscite
or referendum was meaningless.

Soon after his release from jail, Beg went to Delhi to have a meeting with the Home Minister, Pandit Govind Ballabh Pant to convince
him that civil liberties were completely crushed under the Bakshi
regime. Pant disagreed and so Beg on his return to the Valley founded
on 9 August 1955 a new political party—the Plebiscite Front—which
declared that the accession of J & K State to India was purely temporary, and to determine its final status, a plebiscite should be held.
Mridula differed from Beg and she was unhappy at the formation of
the Plebiscite Front. But her presumption that Sheikh shared her
opinion in this case was not correct. The lengthy note the Sheikh
wrote after his meeting with Dr Subbarayan shows clearly that he
regarded plebiscite as the only possible solution. He said that Nehru
had many a time made commitments in public and in parliament that
the people of Kashmir must decide their future in a free and
democratic manner.

The Plebiscite Front was not very active at the beginning. In fact, there were two groups in it—one was the pro-Pakistan group and the other was a non-committed group which wanted Sheikh Abdullah to be released first. The pro-Pakistan group was led by Kochak while the other group by Hamdani, Soofi and others. Though Mirza Mohammed Afzal Beg was deeply involved with the Plebiscite Front, he continued to keep close contact with the Indian leaders. Since his release, he did not attend the social and official functions organized by the government of Jammu and Kashmir State and Karan Singh. But, when the Rashtrapati, Dr. Rajendra Prasad visited the state, Beg and his colleagues welcomed him and offered all possible cooperation. When he was seen at the party given in honour of the Rashtrapati, quite a number of friends, some holding responsible posts in the government, remarked, 'Since your release you have never attended any government parties; how is it that you are here today?'. Beg is reported to have replied, 'This is the Rashtrapati's party. He is the Head of the State and above party-politics. I shall continue to participate in such functions. We are proud of him.' A common friend amongst them asked, 'If this is your policy, then why don't you attend the functions of the *Sadr-e-Riyasat*?' Pat came the reply, 'Unfortunately, the *Sadr-e-Riyasat* has not kept himself above party-politics. In the Assembly and elsewhere, he has associated with one party.' Obviously Beg and his colleagues were eager to keep the doors open for negotiations with the centre but show their disapproval of the Bakshi government. To what extent Mridula associated with the Plebiscite Front can be judged from her remarks:

I am going to meet friends—once colleagues in the struggle for freedom but now finding themselves thrown into the opposite camp. As far as I am concerned, I continue to be a colleague of those who are desirous of maintaining the old bond of mutual trust in bonafide of each other and accept the basic fact that everyone of them is a patriot to the core. This will take me to the homes of many such persons who, in and out of season, are made the target of a maligning campaign.

The government of J & K rearrested Beg on 19th November 1955 as his political activities were considered prejudicial and unlawful. He went on a hunger strike to secure humane treatment for the political prisoners and detenues. Mridula took up their case and suggested that the same treatment should be meted out to them as to prisoners in other states of India. Apart from this, she also took up the cause of the family members of the detenues, specially the issue of allowances due to them. Her house in Delhi became a centre for the activities of some Plebiscite Front leaders who took full advantage of her hospitality and generosity.

Being unsuccessful in her efforts to get Abdullah and his associates released, Mridula renewed her tirade against the Bakshi government. The first target of her attack was the J & K Ingress (Permit) Rules, 1955 introduced by the Bakshi government. She issued a circular letter to all editors of newspapers, in which she argued that the Act, which was meant to safeguard the State from external dangers, was being used for party politics. Every time State subjects wanted to go to India, they had to secure a permit and on numerous occasions they were denied this. Indians, including the subjects of the J & K state, were treated at par with foreigners. She pleaded that there should be a separate system of entry for Indians into J & K and no permits should be required for citizens of J & K.

Although more than two years and ten months had elapsed since the detention of Sheikh Abdullah, there were no signs of his trial beginning. Among those who felt that justice was being denied to Abdullah was Dr P. Subbarayan, a common friend of Nehru and Abdullah. Subbarayan got permission from Nehru to meet the Sheikh in jail on 20 June 1956. For three and a half hours they had free and frank discussions. Abdullah kept a report of this discussion in which he recapitulated in detail the happenings in Kashmir since 1947. Subbarayan suggested that the Sheikh should write to Panditji and seek an interview with him but the latter saw no point in it. He then suggested that in order to facilitate his release, he should issue a statement saying 'My heart is still with India and ... if released, I will try to retrieve the position, though I still hold that the right of self-determination of the people of Kashmir must be respected', but Abdullah was unwilling to do so. A duly signed carbon copy of the above-mentioned recorded conversation was sent by Abdullah to Mridula.[24] He wanted to keep her informed of the developments during his detention and may have wanted her to show the report to Nehru. The story of his dismissal as unfolded by him in his note convinced her even more that the responsibility of getting him ousted rested primarily on Bakshi Ghulam Mohammed.

The state government did not want Abdullah released and to participate in the state Constituent Assembly, which had to finalize the constitution. The Sheikh wrote a long letter to the President of the Constituent Assembly in which he accused the government of having murdered democracy on 9 August 1953 by illegally and unconstitutionally removing him from premiership and arresting him. He said:

Civil liberties in the State have been buried deep, legitimate political activities have been crippled and public life paralysed. Huge amounts borrowed from India are being utilised in corrupting people, granting them contracts and other perquisites in order to prop up your own regime.[26]

Ghulam Mohammad Sadiq, the President of the Constituent Assembly, refuted the charges levelled against him and Bakshi Ghulam Mohammed. He accused Sheikh Abdullah of having deliberately delayed the adoption of the Constitution in 1952–3 and of a 'fascination for independence of a truncated State which would, more or less, include only the Valley of Kashmir'. This, he said, led to a difference of opinion within the Cabinet and 'a parting of ways could not be avoided'.[26] Sheikh Abdullah replied to this by contradicting Sadiq's allegations and once again asserted that they had 'no right to foist a constitution on the people'.[27]

Sheikh Abdullah genuinely believed that the Kashmir tangle could not be solved by ignoring him, as the Kashmiris regarded him as their undisputed leader. It was precisely for that reason that he did not want to issue the statement suggested to him by Dr Subbarayan. The statement made by Jawaharlal Nehru in Parliament on March 29th, 1956 did not bring any change in his thinking about the Kashmir problem. In that statement the Prime Minister reviewing the Kashmir issue, said:

The point now to remember is that the first thing that was required by the U.N. Commission was for Pakistan to withdraw its armed forces from the area of the State occupied by it. There was a great deal of talk about plebiscite as to what India should and should not do. But throughout, the first demand of the United Nations has been the withdrawal of Pakistan forces from the area occupied by them. Other factors come later. We were asked to withdraw the bulk of our forces on Pakistan withdrawing from the area in order to relieve tension, but to retain our army in the State in order to give it protection. *The right of our army to be there was recognised*, but it was stated that since Pakistan is withdrawing completely from Jammu and Kashmir State, India also can reduce her forces as that would tend to bring about a better atmosphere. Today, eight and a half years after invasion, the armed forces of Pakistan are still there. *All this talk of plebiscite and everything is therefore completely beside the point.* Those questions can arise only when Pakistan has taken a certain step, i.e. the withdrawal of its armed forces. Pakistan is out of court till it performs its primary duty by getting out of that part of Jammu and Kashmir State on which it committed aggression. This is a major fact to be remembered. It has been found that the Government of India and the Government of Jammu and Kashmir State could not remain continually in a state of suspended animation in regard to Kashmir; something had to be done. Years passed and certain steps were taken by the Jammu and Kashmir Government, with the concurrence of the Government of India, to elect and convene a Constituent Assembly. We made it clear that while the Constituent Assembly was free to decide on any constitution it liked, we also continued to be bound by our international commitment. Eight or nine years have passed and major changes have taken place and the people of Kashmir have been settled. The President of Pakistan and others repeatedly talked about the 'abject slavery' of the people of Jammu and Kashmir State under the present government. I really do not know why they should talk in this irresponsible manner. Jammu and Kashmir is not a closed book. Fifty thousand tourists went there last year and if there is one thing which is well established, *it is that the State has never been so prosperous before.* It is not for me to say what the state of the people is on the other side of the Cease-Fire-Line. But I notice that there is a continuous attempt by people on that side to come over to this side to share in the prosperity. When we were discussing various ways of resolving our differences with the prime minister of Pakistan, there was a new development. This was the promise of military aid from the United States to Pakistan which was subsequently fulfilled. *This created not only a new military situation but a new political situation.* In discussing this question of Kashmir with Pakistan representatives and others, apart from legal and constitutional issues, we have had the practical aspect in mind; i.e. we wanted to promote happiness and freedom of the people of Kashmir and avoid any step being taken which would disrupt and upset things which have settled down and which might lead to the migration of people this way or that way. Meanwhile, constitutional developments have taken place both in our Constitution and that of the Jammu and Kashmir State. We have laid down in our Constitution that we could not agree to any change in regard to Jammu and Kashmir State without the concurrence of the Jammu and Kashmir Constituent Assembly. In consequence of all these factors, I have made it quite clear to the Pakistan representatives that while I am prepared to discuss any aspect of this question, if they want to be realistic, they must accept the changes and take into consideration all that has happened during those seven and eight years ago. They did not quite accept that position and there the matter ended. The only alternative, I said, was the continuing deadlock in our talks.

Having had nine years of this Kashmir affair in changing phases which has affected the people of Jammu and Kashmir State and of India in a variety of ways, affecting our Constitution, sovereignty and vital interest, am I to be expected to agree to some outside authority becoming an arbitrator in this matter? No country can agree to this kind of disposal of vital issues. But I do think that since both Pakistan and we agreed that on no account we should go to war with each other, we should settle our problems peacefully though they may not be settled for some time. It is better to have a problem pending than to go to war for it.

When Sheikh sahib was denied the facility of attending the Constituent Assembly, Mridula reacted sharply and pointed out:

Constitution-making is not the privilege of one group ... every segment of society should be represented in the framing of the Constitution... Hence, my humble request is that the Constitution should not be finalised unless Sheikh Abdullah and the other members of the Constituent Assembly are released... and normalcy is restored...

By October 1956, the Constituent Assembly decided upon a Constitution for the state which came formally into operation on 26 January 1957. It was modelled on the Indian Constitution and declared that 'the State of J & K shall be an integral part of the Union of India'.[28] Despite protests by Sheikh Abdullah from jail and by the Security Council of the U.N., the new Constitution duly came into effect.

While delivering a speech on the Kashmir issue in the Security Council on January 16, 1957, the foreign minister of Pakistan, Malik Feroze Khan Noon accused India of suppressing civil liberties in the state of Jammu and Kashmir and demanded that India's colonial rule be ended immediately. To substantiate his views, he quoted extracts from various circular letters issued by Mridula to the members of the Indian parliament and the Indian press. This speech of Feroze Khan Noon created a furore in the Indian parliament and among mombers of the public. As a result, Mridula's public image suffered grievously. Unmindful of the adverse public reaction, she continued her efforts for the release of Sheikh Abdullah and his associates. At the same time, she protested in a cable to the President of the Security Council against the speech of the foreign minister of Pakistan for mutilating and distorting her views:

Surprised to read Malik Noon's reference to some of my letters to Indian Parliament members in his speech to Council on sixteenth (.) Saw extracts Appendix 'A' in Pakistan newspapers received New Delhi twentieth (.) Strongly protest for quotations taken out of context misrepresentation distortion of my views (.) Shall be obliged await representation before taking note of extracts presented by Mr Noon (.)

The representation submitted by Mridula to the Security Council is summarized in the following text:

Malik Feroze Khan Noon has tried to make out that (a) from the very beginning, the Kashmiris wanted to opt for Pakistan and the Indian authorities obstructed them from doing so. Since then, they are subjected to terrorism and repression. Steps should, therefore, be taken to save them; (b) India's claim that things have settled down in Kashmir, that the people are reconciled to everything and that law and order prevailed in the Indian occupied area of the State, is false. The facts are otherwise; (c) The people are ripe for revolution if the United Nations do not deliver them from the colonial domination of India.

To substantiate his charges against India, extracts out of context from the said letters have been taken. Not only have I been quoted but words and sentences, not existing in the originals, have been inserted. This mutilation dangerously misrepresents the objective behind the present agitation on the domestic matters. It is diametrically opposite to what we are trying to achieve. And on this basis, he seeks intervention of the U.N.O. If his representation and proposals are accepted, they are definitely going to be disastrous to the interest of the State. So the need for this representation. If Mr Noon wanted to quote me then, in fairness to the members of the Security Council and me, he should have placed my documents in full and allowed them to judge for themselves whether the causes of psychological unrest in the State can by any stretch of imagination lead to his conclusions.

As an associate worker of the National Conference of Jammu and Kashmir State (prior to August 9, 1953) for years I have had the privilege of being in close contact with those parts of Jammu and Kashmir State which are inhabited by majority Muslim population. The spirit of mutual confidence, inter-reliance and desire to be co-builders of new India bind us together. Not once have I found the popular leaders and the majority of the people inclined towards Pakistan. They have not accepted the theory referred to by Mr Noon that in the case of Jammu and Kashmir..., the State territory was contiguous to Pakistan. Political, economic, strategic, cultural, geographical and other considerations, all made accession to Pakistan the natural course.' The Kashmiris are strongly averse to it even today. The leaders of Pakistan are aware of this attitude, yet to gain their ends they have chosen to adopt the strategy of psychological warfare.

Unfortunately, the events of the 9 August, 1953 overtook us. It is a temporary set-back to our national effort. But this domestic dispute encouraged the outsiders to subject us to cold war. My efforts to get the Indian leaders to intervene should be wrongly used by Pakistan's Foreign Minister to hide their own faults and give it an internationsl twist, is a matter of regret.

The U.N. Council's Charter provides for... 'to affirm faith in fundamental human rights, in the dignity and worth of human person, in the equal rights of men and women and of nations, large and small....' It is hoped that its aim to give protection to such countries not by interfering in their domestic matters but by seeing that during their growth they are not subjected to various categories of warfare which includes psychological warfare. Kashmir needs this hedging, not by reopening of the sores or inviting outsiders to protect their interest and thus deprive them of their sovereignty, as suggested by Mr Noon, but by staying the hands of all power-blocks and Pakistan from creating situations aiming at unnerving and coercing the populations and leaders into submission to the cult of two-nation theory. In the interest of truth and fairness, may I request that this document be circulated to other members of the Council.

At a news conference on 5 February 1957, Bakshi Ghulam Mohammed, Prime Minister of Jammu and Kashmir State alleged that:

It is true that Sheikh Abdullah had written a letter to the Security Council. I have seen the photostat copy of that letter in Pakistani newspapers, and from that I can say that it was written by Sheikh Abdullah and it was not fake. The letter was smuggled into Pakistan.

As a result of this statement, a general impression was created in India that Sheikh Abdullah might have actually smuggled a letter to the President of the Security Council. Whether the letter alleged to have been written by Sheikh Abdullah to the president of the Security Council was actually genuine or a forged document became the subject of common talk and discussion in India. To wipe out that impression, Mridula issued a circular letter on February 12, 1957 addressed to the 'Moulders of Public Opinion in India' which reads as under:

Through a statement on February 6, I contradicted Bakshi Saheb and reaffirmed my stand that the alleged letter was a fabricated document. Since then, friends have approached me and said, 'There is general confusion in this matter'. The moulders of public opinion are made to feel that Bakshi Saheb, after all, is in authority and has better and more reliable sources. Surely, Pakistan would not have utilised this letter if they were not convinced about its reliability. Would the *Times of India* correspondent in Kashmir also make a mistake? They added, 'There is a deep feeling, a conviction that Sheikh Saheb has always been with Pakistan.' It is difficult to remove this unless he himself comes out and denounces the letter and Pakistani exploitation of him. Therefore, it would be helpful if you take the public into confidence and place before them the data on which you base your stand.

Convinced that Sheikh Abdullah could never have written any such letter to the President of the Security Council, Mridula expressed her views in one of her letters to the Sheikh as under:

We carefully examined this document and came to the conclusion that it was fabricated on account of the following reasons (i) The alleged letter begins with the sentence, 'your excellencies are perhaps aware that I am completing the third year of my incarceration in a detention camp in the State where I have been whisked off as a result of coup d'etat on August 9, 1953'. It means whoever wrote it has done so sometime in June–July 1956, why did it take seven months to reach its destination? (ii) Nowhere in the text of the letter, there is any indication of inclination towards Pakistan. Why is it then that the sender of the alleged letter surrendered it to Pakistan authorities instead of sending it directly to the President of the Security Council? (iii) The language and the words used in the letter do not corroborate with the style of your writing. (iv) How should one consider it a coincidence that the letter meant for the Security Council should first be given for publication to *New York Times*, just at the time when the Security Council was meeting to hear Pakistan's complaint on the Kashmir question? (v) It is observed from the photostat copy of the letter that the handwriting in it does not resemble that of yours. All of us who are acquainted with the style of your handwriting have come to that conclusion independently. However, there was no difference of opinion amongst us whether or not it would be desirable to take action for giving expression to this fabrication without seeking your views. Many of us wanted to adopt the policy of 'wait and see', but I was not for it. Besides, I was not agreeable to get it checked from you as it would have raised a doubt in our minds that such a letter could have been issued by you. It is unfortunate that in the past you have been the victim of suspicion from many nearest friends and from unexpected quarters. I did not wish to be added to the list of such persons. On my part I continued to keep you informed of the steps I intend to take, giving you sufficient time for any suggestion or advice that you liked to send me in the matter. (vi) I am convinced that no useful purpose would be served by you for taking any action in a unilateral manner.

It is evident from the preceding paragraph that Mridula did not have the slightest doubt about the integrity and loyalty of Sheikh Abdullah for India and felt that it would be insulting even to ask him whether the letter, in question, was written by him or not. In fact, she took it upon herself to contradict the propaganda of Bakshi Ghulam Mohammed about the letter alleged to have been written by Sheikh Abdullah to the Security Council.

Throughout the year of 1957, Mridula made efforts to get Sheikh Abdullah and his associates released from jails. To achieve her objective, she tried to defend the Sheikh from all the accusations which were hurled at him by Bakshi Ghulam Mohammed and his administration from time to time. Not only that, she did not miss a single opportunity of denouncing the maladministration of Bakshi Ghulam Mohammed and his cabinet colleagues. As many as thirty long circular letters were issued by her within a period of one year giving the inside story about the administration of the state.

One of the most important political developments in Jammu and Kashmir was the framing of the constitution for the state by the members of the Constituent Assembly. The Constitution came into being on 26 January 1957. Being dissatisfied with the constitution which did not have the blessings of Sheikh Abdullah, Mridula expressed her displeasure in one of her circular letters. After adoption of the Constitution, the authorities of Jammu and Kashmir wanted to hold fresh elections in the state. The National Conference contested all seventy five seats including four reserved seats, the Praja Socialist Party put up twenty five and the Praja Parishad twenty candidates. Apart from the the opposition political parties, many newspapers in the state also expressed doubts about the fairness of the elections. The National Conference had secured an absolute majority by winning thirty eight uncontested seats in a House of seventy five.

While Mridula campaigned relentlessly for the release of Sheikh Abdullah by highlighting the repressive policies of Bakshi Ghulam Mohammed, the latter too did not leave any stone unturned in maligning her. The most glaring example of such an instance was the story circulated by his agents in the open session of the Indian National Congress at Indore which appeared in most of the newspapers in the following form:

The Prime Minister at the annual session of the All India Congress Committee at Indore, turned away Miss Mridula from the dias. Ultimately, Miss Sarabhai had to leave the dias. Once Secretary of the All India Congress Committee and a colleague of Shri Nehru, Miss Mridula could not understand whether Shri Nehru's annoyance was because of his love for discipline or for his disapproval of the latter's attitude to the Kashmir problem. She not only left the A.I.C.C. pandal but left Indore too, it is learnt.

This was, however, contradicted by some newspapers at a later date because of the reply received from the Prime Minister's Secretariat by Chaudhary Mohammed Shafi, M.P. from Jammu and Kashmir State as under:

The Prime Minister has received your letter of January, 9 1957. He desires me to say that the news-item referred to by you is not correct, but he is unable to go about issuing contradictions to every incorrect statement in the press.

Bakshi Ghulam Mohammed, was deeply perturbed by the activities of Mridula who had been advocating the release of Sheikh Abdullah and his colleagues for the past four years. The circular letters relating to the state of affairs in the Kashmir Valley issued by her, produced such as impact on the people of the Kashmir Valley that it became extremely difficult for the state government to decide what to do with her. As a counter-measure, Bakshi Ghulam Mohammed started a virulent propaganda against her as is evident from a letter dated 9 February 1958 received from one of her friends:

During his recent visit to Delhi, Bakshi Ghulam Mohammed not only flung the wildest charges openly against you but has also involved your father Ambalal Sarabhai. He dubbed you both as Pakistani agents and said, 'saboteurs in Kashmir are being helped with money and propaganda material from a well-known "Trust" of an Indian business house, "Ambalal Sarabhai Trust" owned by Ambalal Sarabhai and Mridula Sarabhai. Lakhs of rupees are being spent to carry on the propaganda and to help sabotage against India and Kashmir by Sarabhais. Two Pakistani papers The Dawn, and 'The Pakistan Times' were receiving materials with lakhs of rupees from India. One should have hoped that this money could be spent on the poor children and widows. I am pained that from this city of Delhi, a lot of propaganda is being carried out against us, against Kashmir and the people of Kashmir by Sarabhais.' To add venom to this whispering campaign, Bakshi Saheb's men at Delhi also concocted a story that you have been financing those involved in bomb cases and the Enemy Agent Ordinance. Bakshi Saheb is believed to have left instructions to make this charge one of the main attacks against you before the members of the Parliament. Even though I am convinced of the falsehood of the story, yet please take public into confidence and tell the truth.

Mridula added in a circular-letter:

Since it is a widely felt desire, I take the liberty of laying down a few facts. For the past fifty-five months, Bakshi Saheb's visits to Delhi have been officially sponsored and officially financed. A maligning campaign is let loose by him against the people of the Valley, against those political colleagues whom he threw into Opposition through the treacherous coup of August 9, 1953 and against those persons who in the interest of Indo-Kashmir relationship wanted to resist the onslaught of the Cold War. As I tried persistently to put before the general public the true facts and warned them not to be misled by Bakshi Saheb, his wrath against me is quite understandable. For the last fifty-five months, I have been under his fire. So, for me the recent utterances are not a new thing but the attack on Shri Ambalal Sarabhai who always kept out of politics and about whom he had full knowledge, is quite detestable. ...: I am privileged to have liberal parents who have successfully worked out a democratic pattern of family relationship in personal life and in the economic structure of the family property. Being one of the social revolutionaries and an emancipator of women in his private life, my father treated his daughters on a par with sons and gave them equal opportunities in life. My parents never imposed the colloquial patriarchal pressure on our lives and activities. Economically, we are all individual units and paying taxes to the Government and as such we cooperatively run our homes etc. Even prior to 1947, this was well known to the then authorities and even now it is known to the present authorities. Throughout the national movement, never did anyone try to coerce my parents for my political activities as Bakshi Saheb has tried to do this time. Perhaps, he forgot that Delhi is not under his jurisdiction. Such being the relationship between the children and the parents, the efforts and the tactics to bring pressure on me having failed, Bakshi Saheb, in desperation, should have gone to such a level of 'decency' is no surprise to me. But what is painful is why the comrades connected with the press have not tried to check the facts either from me or from Ambalal Sarabhai and in the public interest should have given the other side of the story also. Since I came to manage my own affairs, I have tried to contribute the maximum to the social public activities.

The Release of Leadership

Rumours started now floating that Sheikh Abdullah was being released and on 8 January 1958 he was indeed released from jail unconditionally. It is said that on his release, the government of Jammu and Kashmir had offered him the facilities of transport in order to undertake the journey from Kud to Srinagar but he declined to accept that offer. Mridula sincerely hoped that the press in India would take an objective view of his release and cooperate in the maintenance of a congenial atmosphere. But that was not so, and by and large, it adopted a line of supporting Bakshi Ghulam Mohammed. A few examples are quoted below:

There is no evidence of triumphal arches waiting to greet him. Nor, unless the appearances are so deceptive, should he spoil for a brass-band reception. This morning a few crackers went off in a corner of a city lane. If that was the measure of his support in this babble of tongues, then it is obvious that the value of Sheikh Abdullah's pre-detention stands, namely, a sovereign Kashmir has fallen well below zero. (The *Hindustan Times*, 10.1.1958).

There is no doubt that the calm atmosphere that obtains in Srinagar even after the news of his release is known to everybody, has been somewhat of a surprise and disappointment to the Sheikh. Last night, a party of journalists went round Srinagar after the Sheikh's release was announced. There was not the slightest trace of excitement anywhere. The idea that he could be a private citizen in his own state without any power is apparently something which he is not yet prepared to reconcile himself with. Perhaps, the Sheikh is bitter both against the Government of India and the Kashmir Government. He would not rest content until the present administration in the Valley is dislodged. (*The Hindu* 10 Jan. 1958)

Mridula was deeply perturbed because of the indifferent attitude of the press in India at the release of Sheikh Abdullah. To apprise the people of India with the prevailing state of affairs in Jammu and

Kashmir State, she issued a circular letter, dated 16 January 1958 which speaks for itself:

With the unconditional release of Sheikh Abdullah, there is a general feeling of relief in the country that the unhappy chapter of August 1953 episode is over. Sheikh Sahib should now avail of the opportunity and extend his hand of cooperation to revive the national goodwill and understanding between the people of the State and the people of the rest of India. Indeed, the release can be a step towards the basic change in the policy. Nevertheless, the situation seems to be drifting in a different direction. Lots of obstacles are being placed in the way of Sheikh Sahib and efforts are being made to destroy the good results that were expected from his release. Sheikh Sahib's close colleagues, such as Mirza Afzal Beg, Khwaja Mubarak Shah, Hakim Habibullah, Haji Mohammad Ishaq and a few others are still in detention. Many of his friends have not been given permits to go back to the state from Delhi. Some well-wishers of other states in India who are desirous of meeting Sheikh Sabib are also not issued with permits. I am also denied this facility. Once in winter of 1955, the ban for my entry into the State was removed and I stayed in the Valley for about six weeks. Thereafter, the ban was imposed again without letting me know the grounds and it still continues.

The newspapers in India were giving different kinds of news about the speeches of Sheikh Abdullah since his release from detention. Their information was based on the news received from their reporters who were following the itinerary of the Sheikh from Kud to Srinagar. Most of these news reports were at variance with each other and misleading. To apprise the people of India with the true facts about the state of affairs in the state, Mridula gave full details about the Sheikh and his speeches in a circular letter, dated 22 January 1958:

After his release from the sub-jail at Kud (Jammu Province) on January 9, 1958, Sheikh Abdullah proceeded to Srinagar via Anant Nag. To welcome him, the leading members of the Plebiscite Front had arrived at Upper Mundale, situated on the Kashmir side of the Banihal Pass. A large crowd had already arrived there from the adjoining villages. It was a quarter past nine when Sheikh Saheb and the party arrived there. Braving the severe cold, the crowd surged forward in thousands to have the glimpse of their beloved leader. The women sang the traditional folk songs of welcome which are sung on such traditional occasions. Sheikh Saheb first went to the tombs of the local martyrs who had sacrificed their lives while demonstrating against his arrest and detention. The night was spent by him in the *Dak bungalow* where the leading citizens and the local leaders had arrived to pay their personal greetings to him. Scenes of warm hand-shakes and embraces were witnessed there till late at night. In the morning the procession started towards Anantnag, situated at the foothill of Banihal. The road was lined with the people who lived in the villages around. On arriving at Larkpura, *Sher-e-Kashmir* addressed a huge gathering:

'My dear brothers and sisters: I exhort you to follow the path of noble life, enriched by experience and belief that evil breeds evil and goodness gives birth to goodness. Fear none but God. Whatever is done with good intentions finally yields the harvest of goodness. Keep your faith in God Almighty firm and unshakable. Truth must ultimately triumph; for truth is the will and the voice of God'.

According to Mridula, having thanked the people for their overwhelming affection, the Sheikh with his procession moved towards Anantnag. The people in their enthusiasm were pushing Sheikh Shaib's jeep which was running not on petrol but on love of people. Anantnag wore the look of *Eid*. There also, he went to the graveyards of the martyrs who were killed in 1953 and offered prayers. Afterwards, he addressed a huge gathering of more than one lakh people. He began with the recitation of *ayats* (verses) from the Holy Koran and said:

I bear no grudge to my Hindu and Sikh bretheren. As ever before, I am their well-wisher and friend. I want them to remember this only that fraud and deceit can never last forever and ultimately it is the truth that triumphs. I assure my Kashmiri pandit brothers that I am even today the same Sheikh Abdullah of 1947. I have not changed at all. In that stormy period when the lives, property and honour, in fact your existence, was at stake, how I with Mohammed Afzal Beg and my other colleagues stood firmly for your safety. In saving your life we were playing with fire, yet we faced all difficulties and succeeded in giving full protection to you. In doing so we did our duty to God and the Holy Prophet. In no way, it was an obligation on you. My God is the same that I had in 1947 and therefore there will never be a change in my behaviour and attitude. I am not campaigning in any manner when I remind you of my stand. I am only stating it to impress upon you the reality that a boat with cargo of repression and terror is bound to capsize one day or other. The principles of human values, brotherhood and protection of minorities and their honour, lives and property are basic and irrevocable as far as I am concerned. I shall remain true to them unto the last.

Panditji is a very great man. I still have the best regards for him. But if an enemy hits with a stone, it does not hurt so much but if a friend hits with a flower, it becomes unbearable. I want justice. Let him act as a judge in this matter and if I am proved to be guilty, I am prepared to undergo punishment. What had the innocent masses of Kashmir done for which they are riddled with bullets and made targets of inhuman atrocities. In this connection, letters, telegrams and representations were sent to Panditji but he remained silent. So far as my duties are concerned, I cannot barter away the rights of the people of the state for the sake of anybody.

The situations which were created as a result of the tribal raid after the Partition of India are well known to everybody. They wanted to drive us as a flock of sheep and goats. In that grave hour of peril, Panditji came to our rescue which we can never forget. When the danger of tribal raid had been averted from Srinagar, then Panditji assured the people at Lal Chowk that the final decision of accession shall be decided by the people themselves and that no effort will be made to take undue advantage of the helplessness of Kashmiris. This was reiterated in the Security Council and Panditji repeated it time and again. Today I want to enquire from Panditji as to what has happened to those promises and assurances. The future of this country can neither be decided in Karachi nor in Delhi but the four million people of the state, irrespective of caste and creed, will do so with their free-will. The Hindus, Muslims and Sikhs were living in their traditional harmony and brotherhood but a few selfish persons sabotaged it.

Sheikh Abdullah reached Srinagar on 13 January 1958. It was *Urs* at the Hazratbal Shrine. The Sheikh had been the president of the *Majlis-e-Auqaf* for more than twenty years which controlled the management of all the shrines in the Valley of Kashmir including that of Hazratbal. He, therefore, availed of the opportunity of visiting the shrine after a lapse of four and a half years. The people of Srinagar wanted him to address them as he used to do in the past. For them, the Sheikh had been a scholar and a reformer in Islamic religion and mode of life. The Sheikh came to the platform and invited the audience to join him in saying *Allah-o-Akbar*. He began his address by reciting certain verses from the Koran and after reciting an eulogy in praise of the Prophet, he addressed the congregation:

At the time of the Propet's death, there were signs of confusion and disruption and some of the tribesmen who had not yet understood Islam, became apostates, some of them refused to abide by the Shariat and paid *Zakat*. People who were making false promises, organized their parties and tried to draw the people into their fold. Man is weak by nature. Human weakness was exploited at that time as it is being exploited today. Had the Muslims followed the teachings of the Koran, they would not have gone astray. Relating the story of his detention, Sheikh Abdullah said, 'Had this been inflicted on me by an enemy, I would not have felt so sad and aggrieved, but as the arrow was shot at me by some of my friends, it hurt me deeply. I was put behind bars by those who had been brought up by me; and who used to declare that, while for the rest of the Muslims, the five fundamentals were the bases of religion, for them a sixth tenet, an unshakable faith in the Sheikh was equally essential. God is my witness that it was my sincere wish that they should prosper. Alas! not only these friends of mine acted against me, but the world

believes that my revered friend, Pandit Jawaharlal Nehru was also made ac-
complice, though we put up a joint fight against the British and worked
together. My relations with him were those of the members of one family.
During the Movement, he helped us a lot and suffered for us. I have every
respect for him in my heart and to save his honour and prestige, I am prepared
to make any sacrifice. But alas! Panditji, the Indian army, CRP, money and
press are being used against those Kashmiri Muslims who were once prepared
to shed their blood for them are now being tortured and plundered today.
Incidents of this kind were related to me and this made me restless. But the
Holy Koran was with me and when I read it with the commentary, anguish
and despair vanished. If you read and follow the Holy Koran, fear, pain and
gloom would never overpower you.

Some of my colleagues say that Kashmir has acceded to India and that the
Assembly has confirmed this decision. No power on earth can change it. I
say, it was by force that this decision was imposed. Some members of the
Assembly were put under arrest and others are harassed. How then, could it
be taken as a people's decision? I am sorry for those and specially for Nehruji
and Menon who used to claim that India is in the right and is following the
principles laid down by Gandhiji. But now they are misleading people and
distorting facts by saying that the people have already decided about the future
of Kashmir and there is no need now to ask their opinion. If people have
decided, then why repression? As a matter of fact, India has been misguided.
It is unfair to say that a decision has been taken unless it is supported by the
people. Without the people's will it is baseless. Self-determination is the
people's right. Our slogan will continue and our struggle will go on till the
slogan does not take the practical shape. (The audience shouted, 'This country
is ours and the future will be decided by us.') I agree with the slogan. This
is our country. It does not belong to Russia, America, Pakistan or India. Bakshi, Sadiq or Sheikh Abdullah are not entitled to decide its future. Each one
of us has just one vote. The future is to be decided by four million of this
country, consisting of men, women, Muslims, Sikhs, Hindus etc. None else
can decide its future. Mr Nehru was requested to make an enquiry into the
charges against me. If the allegations had been proved, I would not only have
gone from door to door, apologising to the Indian citizens but I would have
no right to live. I regret that even though a thief is given an opportunity to
defend himself, I have been deprived of this right. I requested that justice
should be done to me. One the contrary, the accusation that 'Sheikh Sahib is
a traitor', and that 'he has stabbed India in the back' were dinned into the
minds and hearts of my friends and the Indians. I was alleged to have conspired
with foreign countries. I would never have cared, had Bakshi Ghulam
Mohammed said so. He would have done it out of greed for power. But I am
surprised that Mr Krishna Menon spoke in the same strain.

According to Mridula this speech of Sheikh Abdullah at the Hazratbal Shrine in connection with the *Urs* had a salutary effect on the
audience. They became convinced that gross injustice had been done

to him by the state and the central governments in ousting him and putting him under detention. They felt that for the emancipation of the poor people of Kashmir, he had to suffer at the hands of the government. Thus, he began to recapture the leadership of the masses which had been severed for the past four and a half years due to his detention. The authorities in the State and at the Centre were, however, seriously perturbed by his speech which reflected his uncompromising attitude. The second important speech made by Sheikh Abdullah was again at Hazratbal Shrine. A very large number of people had assembled there for the Friday prayers on 17 January 1958. After recitation from the Holy Koran and a poem in praise of the Prophet, he addressed the audience. The third speech of Sheikh Abdullah on 31 January 1958 at the Hazratbal Shrine was almost similar in tone and character to the previous two speeches. His main objective seemed to be to win the sympathy of the Muslim masses of the Kashmir Valley. Accordingly, he used the Kashmiri language for his speeches and rarely switched over to Urdu. From his speeches it as apparent that he was harping upon the same theme. Mridula felt that of all the three speeches delivered by Sheikh Abdullah had been misrepresented by the Indian press. She felt that Bakshi Ghulam Mohammed who had created an impression in India that the speeches delivered by Sheikh Abdullah had revived an atmosphere of insecurity and instability among the local people and had fanned communal feelings in them. But in fact it was just the opposite and during her recent visit to Srinagar from January 25 to January 30, she had an opportunity to meet thousands of people who had come to Srinagar to congratulate him on his release and to have his *'darshan'* and pour out their grievances to him just as Mridula thought, an adolescent child, separated from his mother for a long time, does when he meets her. The Sheikh's psychological handling of their problems, Mridula felt, gave them solace as some of them had been subjected to torture by the Central Reserve Police and other government agents.

On 17 February 1958, Sheikh Abdullah issued a statement to the press in order to acquaint the people of India with the political stand taken by him on the Kashmir issue. He hoped that the adverse propaganda which was being carried on in the Indian press may be abated because of that statement. But on the contrary, the Indian press became even more hostile towards him as he clearly stated in his statement that a plebiscite should be held to determine the will of the people regarding the future political set-up of Jammu and Kashmir State. The Government of Pakistan had also been insisting that a plebiscite be held at the earliest in order to find a solution to the Kashmir issue. The Government of India, on the other hand, had come to hold the view that after a long spell of ten years when the people had settled on both sides of the Cease-Fire Line of the State, it was not necessary to hold a plebiscite. According to it free and fair elections had already taken place on the Indian side of the State under the new Constitution and the people had already expressed their willingness to remain with the Indian Union. Hence, Sheikh Abdullah's insistence on holding a plebiscite led many Indians to believe that he was aligning himself with Pakistan. The Government of India and the government of Jammu and Kashmir became suspicious of his bonafides. As such, a very close watch on his movements and activities was kept.

On Friday 21 February 1958, a big gathering was present at the Hazratbal shrine. To control the crowd, a volunteer organization under the name of Khidmat-e-Khalaq (service of people) was raised by the supporters of Sheikh Abdullah. They had pitched their tents near the shrine. In his post-prayer speech, Sheikh Abdullah drew the attention of the audience to the serious illness of Maulana Abdul Kalam Azad, Union Minister and called upon them to join him in prayers for his early recovery. He left the Hazratbal Shrine at about 4 p.m. Soon after his departure, a band of one hundred workers of the National Conference emerged from Rajbagh shouting provocative slogans and pelting stones on the volunteers of the Khidmat-e-Khalaq which resulted in a clash. The police intervened and about sixty persons were arrested.

Being satisfied with the initial response of the people of Kashmir, Sheikh Abdullah started holding meetings at other towns. On 3 March 1958, he visited Nauhatta Rai Tang, Khanyar, Koolipora, Napara, Dalgate, Maisuma, Karan Nagar, Batmalu and called on the relatives of those persons who had been detained by the government authorities for associating themselves with him. On March 4, he visited many places around the city of Srinagar. The largest gathering was at Rainwari which was largely inhabited by Kashmiri Pandits. The tone and subject of his speech at this place was quite different from the speeches delivered by him at other places. In the circular letter, dated 6 March 1958, Mridula gave a brief account of his speech as under:

There are some people who are trying to create fear in the minds of Hindus and Sikhs by charging me as a communalist, and are creating misunderstanding in their minds. Fortunately, in this 'Country', the majority and minority communities have very good relations with one another. The roots of this relationship are very sharp. It is true that they belong to different religious groups but that does not make any difference. Barring this difference, there is no other difference. Their dresses, food, customs and the way of living is more or less the same. In fact, it is not the communalism which hangs heavily on the hearts of the people but it is the uncertain conditions which have caused the distress. I am sure, as soon as this uncertainty is ended, communalism will die not only in Jammu and Kashmir, but in India and Pakistan also. To solve this dispute, we need peace in the preliminary stage. So, let us create peaceful atmosphere in the country.

As resalt of Abdullah's speeches, the political atmosphere in the entire Valley underwent a sudden change and it was surcharged with rumours. The State authorities got alarmed and as a precautionary measure, they posted police pickets, militia and the Central Reserve Police at various strategic areas in Srinagar. Wireless-fitted vehicles moved about in the streets and intensive patrolling was carried out in the Mirakadal area and the government promulgated Section 50 in the districts of Anantnag and Baramulla. This was all done to curb the activities of Sheikh Abdullah.who after the promulgation of Section 50, did not hold any public meeting. He refrained from attending the Friday prayers at the Hazratbal Shrine and instead, offered prayers in his village mosque.

After restricting the political activities of Sheikh Abdullah under Section 50 of the Security Rules of the State, Bakshi Ghulam Mohammed turned his attention towards the Muslim masses in India to win their sympathies. He attended the All-India Muslim Legislators' Convention at Lucknow where a resolution was passed supporting his position. On his way back from Lucknow he had an informal chat with some Congress M.Ps. in New Delhi and impressed upon them that Abdullah by his post-release activities had proved the need, once again for his detection:

When I used to come to Delhi before his release, the people used to ask me when Sheikh Sahib is to be released? But today these very people ask why he has not been arrested? I want to make it clear that Sheikh Sahib is not a problem for me. I am bypassing his activities but as soon as I feel that his activities are against the interest of the country, I will take immediate action.

Despite the fact that the political situation in Jammu and Kashmir State had deteriorated considerably, Mridula did not lose hope in

bringing about an accord between Sheikh Abdullah and Jawaharlal
Nehru. In a letter to the Sheikh she wrote:

Had expected some guidance from you on what I have sent you directly or
through Bandhuji. Complete silence has baffled me. I do not know what
should I do. Shall be grateful for clear instructions on the following points
(not through an air-packet but through a special messenger).

1.   What steps should be taken to break the ice between you and J.N.
      (Jawaharlal Nehru). We were working for it in the hope that this could
      be achieved before international complications develop; but we have
      not succeeded (so far). Ashok (Ashok Mehta) will also reach there after
      the Graham Report is out. He is definitely going to be there sometime
      between 10th October and 13th October and stay there for some days.
2.   I want your frank opinion about H.N. Misra (an eminent lawyer of
      U.P.). After seeing him, many of us are sceptical about his capabilities.
      Hence, I want to know if any arrangement is to be made from Bombay
      and elsewhere. Last minute arrangements will be very difficult.
3.   What about my request for associating me with the Legal Defence and
      Relief Committee and placing some amount at my disposal?
4.   Bandhuji (Kashyap Bandhu) told me that if you wanted him to stay on
      and work with you, he will not rush to his village, so I am bit concerned
      as to what has gone wrong. Is there any obvious reason for his reluc-
      tance in meeting the families of the friends? Is he doing so under your
      instructions?
5.   What line do you like me to take when the Graham Report is out?

Without waiting for a reply from Sheikh Abdullah, the same day on
13 April 1958, Mridula sent him another communication in which
she tried to impress upon him that Jawaharlal still had a very soft
corner for him. She informed him that the Consultative Committee
of the External Affairs Ministry had been called on 31st March, 1958
which was attended by sixty M.P.s belonging to all political parties
and was presided over by Nehru. During the discussions, three members repeatedly asked him to clarify the position regarding Sheikh
Abdullah's stand and Mridula's activities. He is reported to have
started by saying that Kashmir was a complicated question. Much
had also been said in the press against Sheikh Sahib. Such an approach did not help the solution of the problem. A member suggested
that he should be rearrested. To this, Nehru is said to have replied
that arrest was no solution to the problem. Then a member reminded
him that Sheikh Abdullah was doing many a thing against the government. Nehru is said to have remarked that 'after all Sheikh Abdullah
had been in jail for more than four-and-a-half years, while Bakshi
and his government have been doing many undesirable things, know-

ing well enough that it is against my desire and wish and I strongly detest them.' Then, referring to Mridula, he was asked, 'Sheikh is in Kashmir and Mridula is here. Why has she not been arrested?' He was told about her writings against the Kashmir government. Replying to this, Nehru is said to have told them, 'Kashmir government is also writing against her. After all they are a government and she is an individual. If anyone finds anything objectionable to her writings, they can sue her in the court. We have a democracy and every individual has a right to express his or her opinion orally and in writing individually or collectively'. When a member said this was very objectionable. Nehru replied, 'Court can decide, the government cannot give such importance to an individual in comparison with the government'.

The two communications sent by Mridula produced some effect on Sheikh Abdullah and he sent the following reply to her on 9 April 1958.

Until now I am in receipt of several communications from you to which I have so far not replied. The reason for my silence is obvious. It is the month of Ramzan and I wanted to spend it in absolute quiet and calm. Government of India also helped me in carrying out my resolve by promulgating Section 50 throughout the Valley through their stooge Bakshi Ghulam Mohammed. This action of theirs has crippled all political activities within the Valley. Still one could have used the vehicle of press for the propagation of one's ideas and principles but, unfortunately, this agency also has been completely purchased and brought under control. The result is that the so-called nationalist press in India and within the State has lost its objectivity and independence and what is most regrettable has turned itself into a propaganda machine for whatever views the ruling party in India holds in regard to various matters. In your latest communications, you have asked for some guidance from me to all you have sent me either in writing or verbally through Bandhuji. I am sorry to say that Bandhuji did not tell me anything specific which needed to be replied. As for your communications, they were mere situation reports which needed no comments on my behalf. Of course, in your last communication, you have put certain questions to me and here is the reply to them:

1. You asked me as to what steps should be taken to break the ice between me and Panditji. You will perhaps remember that since my release both I and Maulana Sayeed have been pressing for the creation of a congenial atmosphere within the State as well as in India so that the problem facing us could be discussed in a calm atmosphere. Unfortunately, the response has been negative. Hundreds of innocent political workers are behind the prison bars, suffering various kinds of torture and inconveniences. All kinds of facilities, both legal and otherwise, are denied to them. Section 50 has been promulgated for an indefinite period and

throughout the Valley, a reign of terror prevails. This all is happening with the full connivance and support of the Central Government, and the so-called nationalist press of India.

2.   With regard to Mr Misra, my opinion about him is identical with that of yours. For the time being we do not stand in need of any lawyer from India.

3.   After the wholesale arrest of the first defence committee, no other such committee was formed. Therefore, the question of your association does not arise.

4.   Bandhuji did not express any desire to stay with me when he met me on his return from Delhi. He was rather in a hurry to go to his village to look after his ailing daughter-in-law. Besides this, thanks to the promulgation of Section 50, I have little work to do and his staying with me would have been useless.

5.   About the latest report of Dr Graham, it is not for me to suggest what attitude you should take about it. Since my release I as well as the people of Kashmir, have made it clear beyond any shadow of doubt as to what they stand for and what they are determined to achieve.

Since the release of Sheikh Abdullah, Mridula had been making efforts to arrange a meeting between him and Pandit Nehru. She had thought that such a meeting would narrow down their political differences, and pave a way for settling the dispute which had led to his detention. But the attitude of Sheikh Abdullah towards such a meeting was indifferent. The ovation and the welcome extended to him on his release by the Kashmiri Muslims in the public meetings addressed by him had strengthened his belief that he was the unquestionable leader of the Muslim masses of the state and that the Government of India would have to admit that fact. He, therefore, did not show any eagerness to meet Pandit Nehru. This attitude of Sheikh Abdullah baffled and hurt Mridula. However, her letter of 3 April 1958 in which she emphasized that Pandit Nehru still held him in high esteem touched him and he communicated to Pandit Nehru on 11 April 1958, as under:

As you know well, I have never indulged in intrigues and falsehoods. There is no doubt that I do not agree with the policy that the Government is pursuing regarding Kashmir. I have made no secret of my views vis-a-vis my stand. I believe that the only fair and just way of ending this ten-year old dispute is by conceding the right of self-determination to the people of Kashmir which you once so ardently supported and for the protection of which the Government of India sent her armed forces into the State in 1947. I am determined to secure this right for the people of Kashmir through all legitimate and peaceful means whatever the cost.

Inspite of all that has happened since August 1953, I still believe that the key to the solution lies in your hands and I appeal to you not to be deceived by Bakshi Ghulam Mohammed and his other supporters in pursuing a policy which, in the end, is bound to prove disastrous for all. I hope you are well.

It was perhaps not prudent on the part of Sheikh Abdullah to have written such an unconciliatory letter to Pandit Nehru. Because of the rigid stand taken by him on the Kashmir issue, an anti-Sheikh propaganda was again started by the Indian press. Besides, his speeches and activities endangered the peace and tranquility of the State and the situation was getting out of the control. Hence, the government had no alternative but to arrest him. He was rearrested on 30 April 1958 and once again taken to Tara Niwas Jail at Kud. He sent a wire to Begum Abdullah about his safe arrival in which he wrote, 'May God protect you all'.

A split had taken place in the National Conference, and a group called the 'Democratic National Conference', with G.M. Sadiq as its chairman had formed a new party. The majority of the members, however, remained with the National Conference headed by the Prime Minister Bakshi Ghulam Mohammed.

Soon after the rearrest of Sheikh Abdullah, the trial of the detenues connected with the Hazratbal Shrine Case started at Srinagar with all seriousness. A committee by the name of 'Sher-e-Kashmir Relief and Legal Defence Committee' had been formed on 7 March 1958 with the blessings of Sheikh Abhdullah. Although a large number of its members had been rounded up, it functioned quite effectively. A complete 'challan' in respect of the accused in the Hazratbal Shrine Case was presented by the police before the Special Magistrate Ganjoo at the Central Jail in Srinagar. Eighty-two persons, including three former members of Parliament stood charged for murder, rioting etc. One of the accused was Maulana Mohammed Sayeed Masoodi, a former Member of Parliament. To unnerve the victims, their relatives, friends and the public in general, the ruling party let loose a whispering campaign as under:

There are many documentary evidences to prove that lakhs of rupees were placed by Pakistan at the disposal of important Sheikhites. To get this confirmed, the names of those persons are being sent to the Union Governments' Interrogation Centre for being subjected to third degree tortures. Thus, the cases have been made fool-proof to enable the victims to be sentenced for fifteen years, if not more. Some might also be hanged.

A suit was filed against Beg and some others accused by the government of J & K. This was known popularly as the Kashmir Conspiracy Case and Mridula was also involved. Her house was searched by the police on 12th June 1958 for any prejudicial literature in the form of news-sheets, pamphlets, posters, photographs, etc., from 10 a.m. to 7.30 p.m. and the search was repeated the next day. A number of documents and her account books were seized. This was followed by disciplinary action against her by the Gujarat Congress Committee. 'A whispering campaign has also been started against those who have dared to keep social contacts with me and in case they continue to do so, they must be prepared to be victimised. Thus an effort is being made to ostracize me by this weapon of law.'[29] She was naturally deeply distressed and hurt by all this and wrote how from August 1946 she had been working for communal harmony and national integration in Noakhali, Bihar and elsewhere. She had been trying to follow Gandhiji's teachings that 'trust begets trust'. It was indeed ironical, she wrote, that 'I am being treated as an accused and the anti-national Bakshi government as the saviour of the nation!'

Mridula Detained

On the rearrest of Sheikh Abdullah on 30 April 1958, Mridula severely criticized the action of the Government of India as well as that of the Jammu and Kashmir government, in particular, Bakshi Ghulam Mohammed, prime minister of Jammu and Kashmir State who according to her, had ruthlessly suppressed civil liberties in the State. She, therefore, started a relentless propaganda against him. He protested to the Government of India for being extremely lenient to Mridula in allowing her to malign the government of his state from New Delhi which fell beyond the jurisdiction of his state. The Government of India had to bow before his pressure and in order to prevent her from indulging in further anti-Kashmir state propaganda, felt compelled to arrest her. An order of detention was served on her under the Preventive Detention Act by the Joint Secretary, Ministry of Home Affairs, New Delhi on 6 August 1958, the text of which is quoted below:

Whereas the Central Government is satisfied with respect to the person known as Shrimati Mridula at present ordinarily residing at B1/48, Diplomatic Enclave, Kautilya Marg, New Delhi, that with a view to preventing her from acting in any manner prejudicial to the relations of India with foreign powers and to the security of India, it is necessary to make an order directing that the said Shrimati Mridula be detained. Now, therefore, in exercise of the powers conferred by sub-clause (a) of Section 3 of the Preventive Detention Act 1950 (4 of 1950), the Central Government hereby directs that the said Mridula be detained.

In pursuance of that order, she was detained in the District Jail, Delhi, popularly known as Tihar Jail on 8 August 1958 and treated as a Class I detenue. The grounds for detention were communicated to her on 8 August 1958 as under:

(a) (i) That you are engaged in carrying on an intensive propaganda against the Government of Jammu and Kashmir established by law and against the administration of the State by the said Government and by the Government of India, in a manner calculated to bring into hatred and contempt the Government of the State and the Government of India and the authorities and persons associated with the administration of that State.

(ii) That you are engaged in carrying on an intensive propaganda advocating the cause of Sheikh Abdullah and his followers who have been endeavouring with the aid and support of Pakistan agencies to overthrow the Government of Jammu and Kashmir established by law and to undermine and delay the integration of the State of Jammu and Kashmir with India.

(iii) That in furtherance of your said objectives, you have been issuing circulars in large numbers, in different languages and at various places, containing false and misleading reports and information, inter-alia about the administration of the State of Jammu and Kashmir by the Government of that State and by the Government of India, about the policy of the Government of India in relation to that State, about the activities of the various authorities of the Government of India in and in relation to that State, about the conditions in that State and about the intentions and activities of Sheikh Abdullah and his followers. Copies of some of the specimen circulars issued by you in the English language containing prejudicial passages are annexed hereto.

(iv) That the circulars issued by you were issued and distributed widely and made available in Pakistan as well as published by the Pakistan press and the Pakistan radio which is calculated to be prejudicial and is prejudicial to India and her cause in relation to the State of Jammu and Kashmir. Some of the reports of the Pakistan press and the Pakistan radio on your circulars are annexed hereto.

(v) That the cumulative effect of the propaganda carried on by you is prejudicial to the relations of India with foreign powers in regard to the cause of India with respect to the State of Jammu and Kashmir as well as to the security of India.

(vi) That you are in regular touch and closely associated with several persons who are hostile to the cause of India in relation to the State of Jammu and Kashmir and are engaged in activities prejudicial to the security of India, including certain persons involved in a conspiracy to overthrow the Government of Jammu and Kashmir established by law and to facilitate wrongful annexation of that State by Pakistan.

(vii)   That you are assisting the aforesaid persons financially and actively in their activities which are prejudicial to the security of India.

(viii) That your aforesaid activities are calculated to prejudice the relations of India with foreign powers and also to prejudice the security of India.

The grounds of detention communicated to Mridula by the central government did not convince her. She was not prepared to admit that the role being played by her in the political affairs of Jammu and Kashmir State was in any way anti-national or detrimental to the interest of the state or to the central government. So on 2 September 1958, she submitted a representation to the central government through the superintendent of Tihar Jail, New Delhi against her detention. In her representation she stated:

Most emphatically I state that as far as I am concerned, I have always taken the stand that the territorial accession of Jammu and Kashmir State to India was complete in 1950. As a 'National and emotional integration worker', working in the State of Jammu and Kashmir, I take it as an accepted fact that Jammu and Kashmir is a part and parcel of the Union of India. The outstanding questions are only relating to the removal of communal prejudices, constitutional adjustments, cutting through the cold war and finding out a peaceful way of return of even that part of the State of Jammu and Kashmir which is on the Pakistan side of the Cease-Fire Line. All that I have advocated is that the people of Jammu and Kashmir be given the same status, rights and freedom as have been given to the people of the other States in the Indian Union, and that they be trusted and a solution be found for ending their troubles in the same manner as we solve problems confronting other frontier regions. This would imply the administration of that State by their true leaders so that the people may not be alienated from India. In other words, I have pleaded and worked for the great cause I have always believed in 'emotional integration'.

I have also openly supported Sheikh Abdullah's public and private appeals that if he or any of his associates including myself were believed to be guilty of any of the offences, there should be an open and fair trial in a proper court so that any prejudice that may be existing may be removed and true facts may come out after proper open judicial trial.

I have tried to live unto and carry on Gandhiji's last mission of 'do and die'. While Gandhiji was advocating the theory of 'Hindu-Muslim Unity or Die' in 1947, he said:

It is on the Kashmir soil that Islam and Hinduism are being weighed. If both pull their weight correctly and in the same direction, the chief actors will cover themselves with glory and nothing can move them from their joint credit. My sole hope and prayer is that Kashmir could become a beacon light to this benighted sub-continent.

The Advisory Board constituted by the Government of India to look into the case of Mridula comprised of two judges of the Supreme Court and a judge of the Sessions Court. She was offered an opportunity to appear before the Board on 20 September 1958. She, however, submitted a written appeal to them a day earlier so that they could understand her stand on the Kashmir issue more explicitly. The complaint and appeal submitted by Mridula had no effect on the Advisory Board. Instead of exonerating her, they recommended her detention for one full year. An order to this effect was issued by the joint secretary to the Government of India in the Ministry of Home Affairs on 24th September, 1958 and said that Smt. Mridula shall continue in detention unto the 5th August 1959.

Unable to get justice from the Advisory Board, Mridula knocked on the door of the Supreme Court. She filed a writ-petition to the Hon'ble Shri S.R. Das, Chief Justice of India and his companion justices of the Supreme Court on 9th October, 1958 against her detention in the District Jail, Delhi. Nani Palkhivala, the eminent counsel appeared on her behalf and contended that her detention was 'malafide', that charges against her had no foundation in fact and the grounds of her detention were vague and had no relation to the object sought to be advocated. The Supreme Court of India did not uphold the conduct of Mridula for the role played by her in Kashmir. Accordingly, the writ-petition submitted by her was dismissed. The *Indian Express*, Delhi, in its edition of 25th November, 1958 reported this news as under:

The Chief Justice observed that each item of the acts alleged had to be judged in the light of the struggle going on between two countries for Kashmir. The Court could not shut its eyes to the facts. Shorn of verbiage the entire argument amounted to 'however mischievous my activities, I must be free'.

The Court felt disinclined to test the validity of the law of preventive detention on the touchstone of the fundamental right of speech and free expression.

The same day, that is 25th November, 1958, she sent a letter of the Chief Justice of India in which she wrote:

I hope to be excused for stating that in my case at least, I did not desire to seek redress on fundamental rights and to get liberty under the condition, 'however mischievous my activities, I must be free'. I would be the last person to be instrumental in undermining the real security of the country and if today the facts of my detention under the given reasons and if the cases in Kashmir against Sheikh Sahib and the Sheikhites based on misleading data, were not resulting in strengthening the real saboteurs, I would have borne the cross without murmur. Sir, as stated earlier in this letter, I can assure you that a struggle between the two countries on Kashmir, is not only a predominant factor for us but also for the people of the State. Therefore, even in greatest provocation, a non-agitation way has been followed by us.

Mridula's detention under the Preventive Detention Act, duly attested by the Supreme Court, made many who had known her since the days of the freedom struggle and had never doubted her patriotic fervour unhappy. They could not believe that she could indulge in anti-national activities which could be against the interest of the nation. Moreover, to keep her behind bars without a trial appeared to them undemocratic and unconstitutional. Acharya Kripalani, leader of the Praja Socialist Party in parliament, was one of them, and he protested against her detention without a trial:

I cannot understand a former General Secretary of the Congress being held under the Preventive Detention Act. Do we hold, like the totalitarians, our best men also under suspicion? Has Miss Mridula done something treasonable? I am sure she is not engaged in treasonable activities. She would not betray the country. Let her be tried. Let there be a fair trial; otherwise, I am afraid, there will always be an impression in the minds of the people that she was engaged in treasonable activities against the Government.

He appealed to the Prime Minister and to the Home Minister that she should be either tried or released. 'Her detention does us no good. Such things happen in Russia. But they should not happen here. There must be some people who are above suspicion of having traitorous propensities towards the country and the people of India.'[30] This prompted Nehru to reply that he had known Mridula

... 'for a trifle over forty years, since she was a child, a girl. And there are few persons in India, men or women, whose courage I have admired so much as hers. She is a brave, courageous young woman. But there are also few persons in India whose judgement I have disputed and thought wrong, often enough. I mean wrong judgement is alike to courage. It often produces very wrong results because she has courage to go wrong, and repeatedly wrong, and not to be cowed down by anybody into any other course of action. But I have been amazed to see how she can persist in wrong doing and harmful-doing to her country. Almost every member of this House has received, no doubt, vast bundles of paper from her frequently. It is amazing—the type of propaganda being carried on and I had that examined repeatedly—much of it is baseless, without foundation. I do not say she deliberately tells a lie; but she believes every liar that comes to her and puts it across to the people with her own imprint and gives publicity to that. We talked to her, tried to reason with her and tried to explain to her, but it had no effect.

He had never doubted her motives, her bonafides. He did not think Mridula was guilty of high treason,

but under an unfortunate set of circumstances, her courage and her capacity is being utilized and exploited for wrong and dangerous purposes. She got far greater publicity in Pakistan than in India. This is no argument, I know; but I merely say that her whole activity—not that she meant it—became so anti-national, so harmful to India that it became rather difficult to leave it where it was.

For months and months, he said that the government took no action because of their high regard for her and because of her own courage. He quoted some lines from Shakespeare that came to his head:

And to be wrath with those we love,
Doth work like madness in the brain.[31]

The preceding paragraphs show that the activities in which Mridula indulged in were not considered treasonable by Nehru but anti-national and harmful for India because of their wide publicity in Pakistan. Having read the texts of the debates that took place in the Lok Sabha on 9 December 1958 in which her name figured, she (Mridula) submitted a petition to the Lok Sabha from jail on 13 December 1958 under Rule 167. In that petition, she gave clarification on those issues which she deemed proper to be brought to the notice of the members of the Lok Sabha, as under:

I am grateful to Acharya Kripalaniji for voicing my demand for a trial and also to the Prime Minister for more clarification about my detention. As the reasons given by him basically differ from the reasons given in the order of detention or on the grounds of detention, so in fairness to the Prime Minister, to Acharyaji, to the Lok Sabha and to the Nation if I request to all the concerned for a frank analysis of the Prime Minister's reply and my statement for objective investigation and search for truth, I hope it will not be considered out of place. During the Gandhian era, many of us who were young at that time had a unique personal relationship with most of the old guards of that period and especially with Nehru. In Bapu, Nehru and a few others, we found encouragement for not surrendering the right of independent judgement to anyone. I consider it my good fortune that after the achievement of independence in 1947, though for many of us the paths have become different, yet our social personal relationship continues on the basis 'trust begets trust', not only with those who are in the Congress but also with others who went into political opposition. To some of them, especially to Jawaharlalji and Pantji who have seen me from childhood and as a camp follower, if I continue to appear as an 'immature' girl, it is but natural. I do not mind it as it is a natural phenomena in human relationship between the young and old. My only prayer to them is that what Gandhiji was to you all, at our age of today, they should be the same to us now.

If the authorities were so sure of their stand that I had been 'spreading lies', then why they did not take the opportunity to detain me under the Preventive Detention Act and get a thorough investigation made into the cases which have been dubbed as lies.

It is a matter of regret that the Prime Minister should have stated that 'her (Mridula's) courage and capacity were being utilised and exploited for wrong and dangerous purpose and she got far greater publicity in Pakistan than India.' If he feels that there is someone as an instigator then why has anyone not been frank with me to enable me to help them to reach the truth.

On personal consideration one should have given up working for the people of Kashmir. But to me the service for the nation has a large meaning. It should be for some cause and in that (process) one has to be prepared to get annihilated completely. That is the lesson I learnt at the feet of Bapu and Nehru. In the context of Kashmir, it is the national and emotional integration which can be achieved by the co-operation of people on the basis of 'trust begets trust'.

There was no reply from the Lok Sabha in respect of the petition submitted by Mridula, as the Committee of Petitions of the Lok Sabha directed that it should be filed. Thus she appealed to all the wings of the government—administration, judiciary and legislature—to seek justice but to no avail.

Though Mridula had been in jail many times during British rule, this was her first experience of jail in independent India:

Eleven years after the attainment of Independence, I was detained in Tihar Jail, New Delhi, on 6 August 1958 under the Preventive Detention Act. Being a political prisoner, I was treated 'A' class. But there was no separate ward for political prisoners in that jail. As such, I was lodged in a female ward meant for criminal prisoners for about three and a half months. On many occasions we had to remain inside the barrack from 7 a.m. to 7 p.m. ... I did not find any marked difference between the prisons of English ruled India and the prison of independent India; ... the only significant difference was that I was getting newspapers without censorship.[32]

She started helping women prisoners by teaching them how to record their statements and reply to questions in court to avoid delay in their cases. With the help of the jail authorities and some social workers, she managed to get ten prisoners in the mental ward who had recovered, released. She was shifted to the political ward and kept in a solitary cell for about eight months. Two convicted prisoners and a matron were posted to serve her, 'that was my entire world'. For the first six months, visitors were allowed twice a month for half-an-hour. Thereafter, they were allowed for four to five hours. Ambalalbhai and Saraladevi tried to visit her regularly. During these meetings, an officer of the Intelligence Bureau was present with a tape recorder. Mridula passed her time in reading or strolling round the ward. She got the impression that the superintendent and doctor of the jail were avoiding her, perhaps on instructions from the Home Ministry. The sanitary and electrical fittings of Tihar jail were, not unexpectedly, sub-standard.

In June 1959, Mridula's father fell seriously ill. The government was willing to release her on parole, which implied giving an undertaking that she would not indulge in any unlawful activity. She declined on the ground that she had done nothing unlawful and hence her detention was illegal. She received regularly news about the illness of her father. Under the rules, she was permitted to write one letter a week; she could receive any number of letters duly censored.

She was to be released on 6 August 1959 but on that day the CPI was to launch a movement against the dismissal of the Communist government in Kerala by the centre. The jail authorities were aware of the CPI's plans and must have felt that they might need the political ward for prisoners and so she was released four days earlier. Mridula suspected that her early release may also have been to prevent some members of parliament from giving her a reception.

Her prison experiences prompted her some years later to write a paper jointly with A.C. Sen on 'Prison Conditions in India' which was presented by the latter at the Amnesty Conference at Kyoto in 1970. Mridula was invited but she sent Sen instead, and bore fifty per cent of his travel costs. This paper was judged as the best in the conference.[33]

Mridula took the issue of Abdul Ghaffar Khan's arrest to Amnesty International which declared him 'Prisoner of the Year' in 1963. 'His example symbolized the suffering of upwards of a million people who are in prison for their conscience.... Non-violence has its martyrs,' said, Amesty.[34]

Since her release from detention, Mridula was very keen to meet Vinoba Bhave (popularly known as Sant Vinoba) who was considered a genuine Gandhian and a true exponent of Gandhian ideology. She wanted to consult him and seek his guidance on some specific issues connected with Kashmir affairs. Vinoba Bhave had gained world recognition because of his novel experiment of '*Bhoodan*' in which the owner of the land, of his own volition, was to gift away one-fifth of his land to a landless farmer. To practice and implement the constructive programmes of Mahatma Gandhi, he had set up an institution at Paunar in Madhya Pradesh which was known as *Sarva Seva Sangh*. Under the auspices of that institution, a meeting of the All India Writers' Conference was called at Amritsar on 11 and 12 November 1959 which was to be addressed by Vinobaji. To get in touch with him, Mridula reached Amritsar on November 10. She stayed back in Amritsar till November 15. During the whole day, she used to be with Vinobaji and discuss with him many issues including the Kashmir problem.

Trial and Concensus

In the year 1959, the Government of Jammu and Kashmir extended its Preventive Detention Act for a further period of five years. This had a far-reaching effect. A person who was in jail at the time of the promulgation of this Act had to continue in jail for a further period of five years. Thus, a person could be detained in jail for a period of ten years without any trial. Mridula raised her voice against this 'black act' and considered it her national duty to apprise the people of the new danger. The people who were detained under the Preventive Detention Act were subsequently put on trial either under 'The Kashmir Conspiracy Case' or 'The Hazratbal Riot and Murder Case'. The Kashmir Conspiracy Case was started on May 22, 1958 but it came up for hearing on 9 June 1958. The principal accused in that case were Sheikh Abdullah, Mirza Mohammed Afzal Beg, Khwaja Ali Shah, ex-deputy minister, Sufi Mohammed Akbar, ex-M.P., Pir Mohammed Afzal Makhdoomi, G.M. Shah, Mohiuddin Zargar, G.M. Hamdani, ex-deputy minister, Khwaja Mohammed Amin, Advocate, Pir Ghulam Ghani, Pir Mohammed Maqbool, Mir Mohammed Nazir, Mirza Ghulam Qadir Beg, Mir Ghulam Rasool and Khwaja Ghulam Mohammed Chikan.

Mridula played a very active role in defending the accused involved in the Kashmir Conspiracy Case and the Hazratbal Riot and Murder Case. She initiated the formation of a Committee for the purpose, which was named the Jammu and Kashmir Legal Aid and Defence Committee. Gul Mohammed Shah, the son-in-law of Sheikh Mohammed Abdullah, was appointed its general secretary. Besides engaging lawyers for the defence of all the accused, one of the objectives of the Committee was to keep the people of the State informed of the progress of each trial. For this purpose, they used to prepare resumes of the replies given by the accused leaders in various courts. Such resumes were distributed to the press, members of parliament and legislative assemblies of various states including important members of the political parties in India. The Committee held that in the past, concocted stories about Sheikh Mohammed Abdullah were spread by the ruling party and sections of the press. Hence, to counter-balance their effect, this course was necessary.

The trial of Sheikh Abdullah created further dissension among the people of Kashmir and its progress was so slow that people started losing faith in its fairness. Mridula argued that the case against the Sheikh should be withdrawn unconditionally. At the same time, she continued her efforts to assist him in his trial. To a great extent she was successful in this endeavour. The Legal Aid Committee for Jammu and Kashmir State consisted of some very prominent jurists and politicians of India, such as:

Sir C.P. Ramaswami Aiyar : Former Advocate General of Madras

Sir Sultan Ahmed : Former Judge of Patna High Court, a member of the Viceroy's Executive Council 1941–3

Shri Porus A. Mehta : An advocate who joined the Law Ministry of the Government of India

Shri B.K. Khanna : Former Advocate General of Punjab and senior advocate of the Supreme Court

Shri B. Shiva Rao : Ex-Member of the Parliament and an eminent journalist. Member, Indian Delegation to the U.N. General Assembly Sessions in 1947, 1948, 1949, 1950 and 1952.

Shri A.M. Khwaja : Founder, Member of Jamia Millia, Chancellor and Member, Executive Committee, Aligarh Muslim University.

The Convenor of the Committee was Shri B. Shiva Rao. The following statement was issued by the members of the Committee:

Reports have appeared in the Indian press during the last two years and a half regarding the proceedings in the three cases which are being held in the

State of Jammu and Kashmir, namely the Kashmir Conspiracy Case, the Haz-ratbal Case and the Bomb Case. We are in no way interested in the internal affairs of the State, or in the political views and affiliation of the accused persons in these cases. Our only concern is with the simple but vital issues that an accused person, whatever may be the nature of the charges levelled against him, is entitled under the Constitution to a fair trial in a free atmos-phere and to a full defence. It is from this stand-point that we have considered the position of the accused in these cases. Our Constitution whose provinces extend to Jammu and Kashmir has enshrined in its provisions the principles of justice and fair play. It guarantees to all the citizens of India the rule of law in respect of proceedings affecting their lives and liberties.

The Kashmir Conspiracy Case was committed to the Sessions Court and its proceedings started with effect from 30 April 1962 and continued till 8 April 1964. During that period Mridula who was an associate member of the Legal Aid and Defence Committee and in charge of the legal defence work in New Delhi was served with an order by the government of Jammu and Kashmir not to enter the territory of the State. The defence of the case was led by Mirza Mohammed Afzal Beg who himself was facing trial. The Legal Aid and Defence Committee also engaged a British lawyer, Mr Dingle Foot of international fame who was ably assisted by his junior, T.O. Kellock. Thus, a team of the following lawyers was formed for the defence of the Kashmir Conspiracy Case:

Mirza Mohammed Afzal Beg, Mr Dingle-Foot, Mr T.O. Kellock, Mr Mohammed Latif Qureshi, Kashmir, Khwaja Mubarak Shah, Porus Mehta, Bombay, S.M. Ahmad, Ranchi, Sardar Amar Singh Am-balvi, Chandigarh. The prosecution had a long list of lawyers appearing on their behalf which included their senior counsel, Shri Nageshwar Prasad of Patna and Shri G.S. Pathak of Allahabad.

Dingle-Foot strongly opposed the prosecution's application for amendment of the charges in the Kashmir Conspiracy Case against Sheikh Abdullah and others in order to bring in the offence of con-spiracy to wage war against the State. He argued in the court of the Additional Sessions Judge, M.K. Tikku that the prosecution was trying to bring in an 'entirely new charge against the accused of which they had not given the slightest hint all these years. We cannot have them making these new allegations. It will be highly improper and prejudicial to the interests of the accused to concede their re-quest.'

The presence of Dingle-Foot in India as a defence lawyer for Shiekh Abdullah gave a new dimension to that trial. Inviting a lawyer of such international repute from a far-off country was quite enough to create a stir amongst the intellectuals in India, and it was but natural that some of them began to speculate as to who was behind this move. Some thought that it could be the Government of Pakistan, while the majority thought that the person could be none else but Mridula. They had reason to believe that only she could afford to bear his expenses. Anyway, the trial of Sheikh Abdullah started with all seriousness.

It would be in the fitness of things to quote an extract from the text of the statement made by Sheikh Abdullah in the Sessions Court in reply to the charges framed against him in the Kashmir Conspiracy Case:

We are charged with conspiracy to overawe the lawful government of the Jammu and Kashmir State, a conspiracy alleged to have begun in 1953 and to have continued until 1958. This raises the issue of 'where the lawful government was to be found'. In 1953 I was the Prime Minister of the State. I had the unanimous support of the Legislative Assembly. On the night of the 8/9th of August, I was summarily dismissed by the *Sadar-e-Riyasat*. I was then imprisoned without charge or trial under the Public Security Act/the Preventive Detention Act. This dismissal was wholly unconstitutional and, as I shall submit, unlawful. It follows that the so-called government which was thereinafter installed was improperly appointed and had no lawful authority. On this ground the Prosecution must fail. Even though I have regarded the government as being without lawful authority, I have never sought her destruction by resort to violence. Indeed throughout my political career which extends back more than 30 years, I have been a believer of non-violence—a principle which I learnt from Gandhiji. I have lived unto it throughout all phases of political struggle.

I was alleged to have entered into the conspiracy as early as 1953. If the evidence of the Prosecution is to be believed, they were fully informed of the progress of the conspiracy in 1955. Yet I was unconditionally released on the 8th of January, 1958 and it was not until the 23rd October, 1958 that I was charged with these offences. Except for a period of less than 4 months, I was in detention throughout the period of the alleged conspiracy, heavily guarded by the Army and the Armed Constabulary of the Government of India and thus completely under the lock and key of the prosecutor. This very fact alone negates any possibility of conspiring or executing the conspiracy as alleged.

At every stage the government of the State has harassed the defence and those who are seeking to assist the accused. Two of the counsels for the defence were detained for periods of about a year under the Preventive Detention Act. Miss Mridula who sought to assist the defence here was detained and later on externed from the State in February 1961, and forbidden to re-enter.

The written statement submitted by Sheikh Abdullah to the Sessions
Judge in his defence merited the attention of a good number of intellectuals including the ranking politicians of India. The logic
propounded in that statement deeply impressed them. They began to
feel that the Conspiracy Case levelled by the prosecution against the
Sheikh was very weak, and that it would not be in the interest of
the country to continue his trial when nothing substantial would be
established against him. In the book *My years with Nehru-Kashmir*,
B.N. Mullik, formerly Director of Intelligence Bureau, Government
of India, during 1950–65, writes:

Many times the accused almost non-cooperated with the Court. They even
insulted the magistrate. On the slightest pretext they moved for revision in
the High Court, forcing an adjournment of the case. Unfortunately, the Kashmir magistrate could not take a strong attitude and so the accused had things
very much their own way. But, whilst the accused delayed the proceedings
at every step, their propaganda machinery in Delhi, working under Mridula,
went on alleging that it was the prosecution which was deliberately delaying
the trial because it had really no evidence against Sheikh Abdullah and others
and was, therefore, afraid of exposing its case before the defence, and was
resorting to this tactic in order to hold these accused persons in indefinite
custody on the excuse of a trial. Most of the blame was ascribed to Bakshi
though, of course, Bakshi himself had nothing to do with the conduct of the
prosecution. The propaganda took such a persistent and virulent shape that
even Pandit Nehru was swayed by it and on several occasions he enquired
from me why the case was moving at a snail's pace. I gave convincing replies,
both orally and in writing, mentioning how delay was being caused by the
accused; how many adjournments they had sought for and how they
deliberately wasted time over futile cross examinations. Shri G.S. Pathak had
also told the same thing. Though Pandit Nehru fully understood that the accused themselves were responsible for the delay, yet he was upset at the fact
that delay had taken place and we had been unable to prevent it.

Meanwhile, in October 1962, China invaded India and Sheikh Abdullah offered his services to Nehru in this hour of crisis. Mridula
continued her efforts to get the court cases against Abdullah and his
colleagues withdrawn. She personally contacted a large number of
MPs and was able to win over about fifty of them to her view-point.
As a result, they wrote a letter to Nehru asking for the Sheikh's
unconditional release. Among the signatories were the veteran Congress leader, N.G. Ranga, Buta Singh, Chandra Shekhar, the socialist,
Ram Manohar Lohia, an oldtime freedom fighter and Gandhian-
Indulal Yagnik, trade unionist S.M. Banerjee and Dahyabhai Patel

Sardar Vallabhbhai Patel. Jayaprakash Narayan had also written to the Prime Minister asking him to withdraw the case. The *London Economist* wrote on 5 October 1963:

The fifty Indian opposition MPs who this week urged withdrawal of the charges against Sheikh Abdullah, the former Prime Minister of Jammu and Kashmir, gave voice to a feeling that is gradually gaining strength in India. The start of his trial aroused relatively little interest. But a year of court reports has done its work. After thirteen months, the prosecution was not even half way through the case.

Meanwhile unusual political developments took place in the Congress party. The Congress Working Committee decided to implement the Kamraj Plan according to which all senior members of the Congress above the age of sixty years holding ministerial posts were to resign in order to make way for junior members. Although Bakshi Ghulam Mohammed headed the post under the National Conference, along with other Chief Ministers and Union Cabinet Ministers, he also volunteered to tender his resignation from the premiership of the State. Pandit Nehru accepted his resignation in October 1963 and the choice for his successor fell on G.M. Sadiq who wielded considerable respect in the National Conference. But due to the manoeuvering of Bakshi Ghulam Mohammed, G.M. Sadiq, the senior-most member of the National Conference, was by-passed and Khwaja Shamsuddin was elected as Chief Minister, and Sadiq did not join the Cabinet as he was not willing to work under Shamsuddin.

Shamsuddin had hardly been chief minister for a couple of months when an upsurge rocked the State. The *Moe-e-Muqaddas* or the holy hair of the Prophet was stolen from the Hazratbal Mosque on the night of 26 December 1963. The relic had been brought to Kashmir at the instance of the Mughal Emperor, Aurangzeb (1658–1707) in the eighteenth century. It was kept in a small glass tube and '*deedar*' was ritually exhibited ten times a year; otherwise, it was kept locked away in a wooden cupboard. The theft of the relic infuriated the Kashmiri Muslims who blamed the government for not making adequate arrangements for protecting it. It was widely held that Bakshi Ghulam Mohammed was somehow involved in the outrage and angry mobs set fire to cinema halls and other property belonging to him and his family. Hostility was also directed towards the Government of India. Sensing the gravity of the situation in the Kashmir Valley, Nehru assured the people of Jammu and Kashmir in his broadcast on the night of 1 January 1964 that:

All efforts will be made to trace the relic. The man responsible for the deed was no friend of the country and everything must be done to find him and punish him. To catch the culprit, it is necessary to maintain peace and calm as disturbances will only help him to get away. The head of the Central Intelligence Bureau had been sent to Srinagar to make a thorough investigation.

The relic was ultimately traced largely due to the efforts of the Director, Central Intelligence Bureau and restored in the Hazratbal mosque.

The Government of India was very keen to bring a change in the administration of the State which was only possible if the Chief Minister, Shamsuddin stepped down. For this purpose, Lal Bahadur Shastri was deputed by Prime Minister Nehru. He had an onerous task. To find a successor was not difficult but to get him elected as the leader of the State Assembly was the problem. Khwaja Ghulam Mohammed Sadiq had the reputation of being honest, straight-forward and a man of integrity. But he did not enjoy the support of the majority of the M.L.A.s who belonged to the Bakshi group. With untiring efforts, Shastri canvassed for Sadiq and was able to persuade Bakshi to get him elected as Chief Minister. With reluctance Bakshi did so. On 28 February 1964, Sadiq assumed charge as Chief Minister. He brought some liberal changes in the functioning of the government which brought a favourable comment even from Mridula. He lifted the ban which prohibited her entry to the State of Jammu and Kashmir. Besides, he issued orders for the release of many detenues.

Nehru had been extremely unhappy with the slow pace of the trial of Sheikh Abdullah. The trial had dragged on for more than four years and no sign of its end was visible. This led him to believe that the Government had no cogent reason to detain him any longer. He did not want the Sheikh to languish in prison without valid grounds. Not having received satisfactory replies from the concerned authorities about the trial, he thought that it was not in the interest of the government to continue the trial of Sheikh Abdullah. The new Chief Minister, Khwaja Ghulam Mohammed Sadiq was also eager to release Sheikh Abdullah in order to reduce tension in the State. Consequently, Sheikh Abdullah was released unconditionally on 8 April 1964 by the Additional Sessions Judge. This was a great day for Mridula, as she felt that their stand had been vindicated.

Soon after his release, Sheikh Abdullah proceeded to Srinagar in the midst of great jubilation. A letter written by Jawaharlal inviting him to Delhi was received by him on 7th April. After celebrating *Eid*

with his family, he reached Delhi on 29th April 1964. At the invitation
of Pandit Nehru, he stayed at Teen Murti House, the P.M.'s residence
as his guest till May 4, 1964. He had a Press Conference in the Press
Club where Mridula writes, 'The press was packed in'. On 1 May
1964 he offered his prayer in Jama Masjid and later on delivered a
brief speech:

A person struggling for a righteous cause is not usually discouraged by difficulties. He has the faith that the objectives he has in mind would be achieved
provided every action is motivated by the pleasure of God and service of His
creation. All of us are passing through a difficult period. It is our endeavour
to overcome the present state of affairs. My colleagues and I have come here
to meet the leaders with a view to maintain peace and tranquility in the Indo-
Pak continent....

The main objective of Sheikh Abdullah was to bring an accord between India and Pakistan to solve the Kashmir problem, and for that
purpose he wanted to play the role of a mediator. He held the view
that the only cause of friction between India and Pakistan was the
Kashmir tangle and if that could be resolved to the satisfaction of
both the countries, peace would reign supreme in the sub-continent.
During his Delhi stay he met old friends and colleagues with whom
he exchanged views on political matters. At the instance of Pandit
Nehru, he had long talks with G. Parthasarthy and Y.D. Gundevia,
Foreign Secretary in the Ministry of External Affairs. During these
talks, Sheikh Abdullah was assisted by Mirza Mohammed Afzal Beg
and his son Farooq Abdullah. From Delhi, Sheikh Abdullah proceeded
to Wardha to meet Vinobha Bhave and Jaya Prakash Narayan—the
Sarvodaya leaders and followers of Gandhian philosophy. During the
period of his detention, they had publicly expressed sympathy for
him and denounced the action of the government in keeping him
imprisoned for such a long time. From there he proceeded to Madras
to meet C. Rajagopalachari (Rajaji) who had also publicly expressed
concern about his long detention. On his return to Delhi, Sheikh Abdullah evolved a new formula for the solution of the Kashmir dispute
which was different from the plebiscite formula. He suggested a confederation of India, Pakistan and Kashmir. He discussed the formula
with Nehru who advised him to talk over the matter with his Foreign
Secretary, Gundevia and G. Parthasarthi. Sheikh Abdullah held that
the Kashmir dispute could not be resolved unless there was goodwill
between India and Pakistan. For cultivating that goodwill he stressed
that the minorities in both the countries should be given proper status.

Parthasarthi and Gundevia pointed out to Sheikh Abdullah that he had come to Delhi to explore the possibility of resolving the Kashmir problem and if he felt that Pakistan was an equally important factor then it was essential to find out their line of thinking. Parthasarthy and Gundevia talked to him at length on what had transpired during the six rounds of talks with Pakistan from 26 December 1962 to 16 May 1963 and tried to impress upon him that India had failed to sell to Pakistan the acceptance of the Cease-fire Line with some adjustment. The Sheikh generally agreed that it was one of the practical solutions but he wanted to be armed with some other alternative before proceeding to Pakistan.

Nehru had given clearance to Abdullah to visit Pakistan if the visit could result in the solution of the Kashmir problem. The Sheikh, therefore entered into correspondence with President Ayub Khan of Pakistan. who invited him to Pakistan vide a letter dated 6 May, 1964. On reaching Pakistan on 24 May 1964, Sheikh Abdullah availed of the first opportunity of meeting President Ayub Khan in Rawalpindi. He was given a hero's welcome by the citizens of Rawalpindi on his first arrival there after Partition. The visit to Pakistan was planned for a period of sixteen days. The first round of talks with President Ayub Khan began on a cordial note. But when Sheikh Abdullah proposed a Confederation of India, Pakistan and Kashmir, it was instantly turned down by Ayub Khan. He insisted on a plebiscite under the supervision of the United Nations. As nothing remained to be suggested, the Sheikh requested Ayub Khan to meet Nehru to discuss the matter. The President gave his assent to this. So in a press conference at Rawalpindi on 26 May 1964, Sheikh Abdullah announced that his immediate objective of bringing about a meeting between President Ayub Khan and Pandit Jawaharlal Nehru had been achieved and that he would return to India a happy man.

Unfortunately, Pandit Nehru passed away on 27 May 1964. On hearing the news, Sheikh Abdullah called off the rest of his tour in Pakistan and rushed back to Delhi to join the funeral procession of Pandit Nehru. The idea of the meeting planned by him was not pursued further as he knew that after Nehru's death, there was no Indian leader to sympathize and encourage him in his views about the solution of the Kashmir problem. Bereft of Nehru's sympathy and support, the Sheikh thought it proper to lie low for the time being and watch the developments. On the other hand, the political situation in Jammu and Kashmir State was not stable. Bakshi Ghulam

Mohammed and his group who had fallen from power were trying to consolidate their strength and were creating all possible obstructions for the new Chief Minister, Khwaja Ghulam Mohammed Sadiq. Sadiq wanted to institute an enquiry against Bakshi and his associates, against whom some serious charges of corruption had been levelled. After getting clearance from the central government, G.M. Sadiq got Bakshi Ghulam Mohammad arrested.

After the death of Nehru, Sheikh Abdullah was left with no immediate programme. Having been confined to prison for more than eleven years, he needed some respite before actively plunging into politics. In order to get away from that environment, he thought of visiting Saudi Arabia on *Haj* (pilgrimage). For this purpose, he needed the assistance of Mridula who had direct access to Prime Minister, Lal Bahadur Shastri. For visiting Saudi Arabia and other countries he needed clearance from the Government of India. With the assistance of Mridula, all the formalities were completed and reservations were made for him on the boat 'Mohammadi' which sailed from Bombay on 13 February, 1965.

At that time the invasion of India by China in October 1962 was fresh in the memory of the Indian people. The manner in which China had stabbed India in the back had created extreme bitterness and many Indians looked upon China as an enemy. The meeting of Sheikh Abdullah with the Prime Minister of China, Chou-En-Lai in Algiers, was considered by many Indians as highly treasonable. Mridula Sarabhai had thrown light on this meeting in her report of 7 April 1965 wherein she states that the opinion of the Kashmiri people was based on the broadcasts from the radios of Pakistan, India and Srinagar. As such, the general reaction of the people was that Sheikh sahib had not foreseen the repercussions of this meeting. However, the extraordinary interest shown by the Foreign Minister of Pakistan in arranging the meeting between Chou-En-Lai and Sheikh Abdullah led many Indians to believe that there was a deep conspiracy on the part of Pakistan. The Government of India took the immediate step of impounding his passport and cancelling the names of all the countries that were endorsed on his passport except Saudi Arabia.

On 19 April 1965 an interview of Sheikh Abdullah was released to the press at Jeddah, the capital of Saudi Arabia:

For our part we are determined to recover our independence and our sovereignty be it by respect for our people's will or by force. We shall obtain our complete freedom. It is a natural and divine right recognised by all people of the world.

Mridula promptly issued a statement on 21 April 1965 under the caption 'A Word of Caution' about this news-item:

Sheikh Sahib had never stood for Independent Kashmir. He has held that it was one of the solutions that had often been considered but that it can come into being with the consent and good-will of India and Pakistan. As to taking recourse to the path of force and violence, Sheikh Sahib has always been against it. His stand has been that for the solution of the Kashmir problem, there should be peaceful atmosphere and it has to be through negotiations and persuasion. Hence, we have drawn Sheikh Sahib's attention to this para. I have also drawn attention of the authorities to this ambiguity of the report and suggested to them to make a thorough enquiry before coming to any conclusion.

The Government of India felt that the confidence reposed by it in Sheikh Abdullah had been betrayed by him. He had been issued a passport to go on Haj but he had misused this trip for political purposes. His meetings with Chou-En-Lai and with King Faisal of Saudi Arabia, it was assumed, were for getting support for his stand on Kashmir. The Minister of External Affairs said in the Lok Sabha that Sheikh Abdullah meeting the Chinese Premier and his accepting an invitation to visit China were extremely objectionable. His seeking Chinese support in regard to Kashmir, was a highly prejudicial act. Preparations were made to arrest him as soon as he landed in Delhi. Among the few who came to know of this secret was Mridula. A.C. Sen recalled that Sheikh Adhullah, Mirza Mohammed Afzal Beg and Begum Abdullah arrived at Palam Airport, Delhi on 8 May 1965. They were met at the airport by G.M. Shah, the son-in-law of Sheikh sahib. After meeting them, Shah issued the following statement:

I was allowed a brief interview before Sheikh Sahib and Beg Sahib were taken by the police to Ootacamond for detention. Sheikh Sahib told me to convey to all friends that they should not lose hope and continue to put their best endeavours for achieving Indo-Pak amity. He further told me that Government and public leaders of Great Britain, United Arab Republic (U.A.R.), Algeria, Saudi Arabia and other countries that he visited, fully know that his objective had been to canvass support for bringing about Indo-Pakistan friendship.

The detention of Sheikh Abdullah and Mirza Afzal Beg at Ootacamund and the order served on Begum Abdullah not to enter the State of Jammu and Kashmir created a strong reaction among the people of the Valley. Mass demonstrations were held and there were violent clashes between the police and the people. The police had to resort to firing at some places to control the mob. On the arrest of Sheikh Abdullah, G. Ram Chandran a member of the Rajya Sabha made a speech in the Upper House on 10 May 1965. Mridula availed of the opportunity of circulating that speech for the information of the general public. A few excerpts from it are quoted below:

Then I would only touch upon one thing more within the short time at my disposal. Sheikh Abdullah has been arrested. I have seen today in the newspaper that Acharya Vinoba has expressed his distress that this has happened, and he used the expression that there has been lack of correct judgement in regard to the matter. My friend, J.P. Narayan has said the same thing. Now Sheikh Abdullah came back to India. He was not running away. If he was an enemy of India, if he wanted to remain outside and carry on the kind of nefarious propaganda which is attributed to him, why did he come back to India? I am sorry that Sheikh Abdullah was arrested without being given a chance to say what he wanted to say to our leaders. If we were not satisfied, then we had every right to put him in prison.... If there is any chance, we must befriend this man. I had talks with him twice and he said to me, 'There are three imperative things for a real settlement. India must be satisfied, Kashmir must be satisfied and Pakistan must be satisfied in order that any arrangement finally made is stable.' I think this is absolutely correct. But I am told that it would be a miracle if he could produce a scheme which could satisfy all the three. Let us put the onus on him to produce that thing which can satisfy alike Kashmir, India and Pakistan and if he cannot, he stands self-exposed.

With the arrest of Sheikh Abdullah by the Indian government, Pakistan though that the political situation in Jammu and Kashmir would be in state of flux. For them, this was the most opportune time to invade the State as they thought that the people of the Valley would rise in revolt against the Indian government. In August 1965, they started sending infiltrators into the Kashmir Valley to create chaos and disturbances. They exploded bombs in public places and sabotaged military installations to create panic among the people. The chief U.N. Observer had brought this fact to the notice of the Security Council but no steps were taken to withdraw the invaders. Ultimately, Pakistan declared war on India in September 1965. The war continued for twenty one days. Thereafter, through the intervention of the Security Council, a ceasefire was brought about between the two warring nations. Diplomatic relations which had broken down due to the outbreak of war were restored. But the tension between the two countries continued. During this period, Mridula continued issuing her newsletters and circulars pleading for Abdullah's release. Some of these were leaked out and published in the Pakistan newspaper *Dawn*. She wrote to Dr Rakhakrishnan and Dr Zakir Hussain, the President and Vice-President of India respectively and also met them to explain her position. L.P. Singh, the then Home Secretary to the Government of India, who had known Mridula in Bihar since 1947, met her. They had a frank discussion in which he told her that the facts in the bulletins issued by her were not correct and the government was concerned since they were being sent to foreign embassies. She insisted that her information was correct and it was only because the government was 'blocking out the truth' and misleading the people that it had become necessary for her to issue these bulletins. She was in touch with Amnesty International which expressed concern over Sheikh Abdullah's house arrest in Kodaikanal and Beg's confinement in Srinagar. It said that there were three hundred of Sheikh's supporters in Jammu jail, 'a prison of which the Indians themselves are reputed to be ashamed'.[35]

The Intelligence Bureau, being suspicious of her activities, kept a close surveillance on her and anyone who associated with her. Her servants, friends and colleagues were all questioned. Mridula feared that she was going to be arrested and did not want to be detained in Ahmedabad as it would cause inconvenience and worry to her family. As expected, a detention order was served on 3rd November 1965. She was asked to proceed to Ahmedabad and remain in the 'Retreat' and not associate with anyone without the written permission of the commissioner of police; nor was she to communicate with anyone except through him. She refused to enter the 'Retreat' which was her parent's home since she did not wish to expose her parents and family to police vigilance and after considerable argument, the police agreed to stand near the gate and not follow her inside the compound. During her internment her activities were naturally curbed; her letters were screened by the commissioner of police. Her father had suffered a stroke in 1964 and was not well. He was not told that Mridula was interned in the Retreat and the house was under police surveillance, as it was feared that this might upset him and affect his health. To Rao Saheb Patwardhan, she wrote: 'After *Azadi*, detention and internment through the hands of those one thought belonged to the same ideology was an experience terribly painful and shocking.'[36] Most people avoided her but among the few who sought an interview with her was Minoo Masani and she was very grateful for this. While she was in detention, she received more than three hundred Christmas cards from Amnesty members of different countries. This moved her deeply and she wrote an article on Christmas cards for the Amnesty International Bulletin. She used to correspond with a number of persons of Amnesty International in Denmark, Norway, Sweden, U.K. and Australia.

Mridula was hurt that Lal Bahadur Shastri, Gulzarilal Nanda and others who knew her personally, believed intelligence reports about her without checking with her. Nanda, Y.B. Chavan and Shastri had all served as home ministers and Shastri was then Prime Minister. To Shastri, she wrote, 'You and Nandaji are in government and we are out. Yet we belong to the Gandhian age and so have a responsibility to see that disruptive anti-Gandhian forces do not succeed in bringing about a breach between us.'[37] To Y.B. Chavan, the home minister, she wrote, 'It is now about thirteen months that I have been interned in Ahmedabad under DIR without being informed as to why this punitive action against me has been taken and why have I been subjected to loss of liberty.'[38] She asked Gulzarilal Nanda:

What have I done to justify internment and how did my freedom jeopardize the defence of India and civil defence of the country? ... I have no doubt that whatever I have done, I have done honestly in the interest of the country and its security and as a part of my duties and responsibilities as a *Shanti Sainik.*[39]

Among the few who pleaded for the withdrawal of the DIR against Mridula was Acharya Kripalani who made a powerful appeal in the Lok Sabha. Mridula, he said, had been a companion of Gandhiji, and of Jawaharlal Nehru and was Secretary of the Kasturba Fund:

And what is the charge against her? That her activities are inimical to the country. Now, I can certify that this lady is as patriotic as anyone of us here, whether in the Congress or in the opposition...

She is a disciple now of Acharya Vinoba, and at his instance she is in the non-violence army: the *Shanti Sena.* Her only fault was that she exposed the corruption that exists in the successive governments of Kashmir. It may be that her fault was that what she wrote was quoted by Pakistan. But I say some of our speeches are quoted in Pakistan.

Why are you degrading democracy? Why are you degrading yourself? You must remember the Rowlatt Act. It was never used. Yet it was the Rowlatt Act that began our fight for freedom. Why do we forget our ideals when we are in power? Mridula was my student, I say that with all the confidence I possess... she is a patriotic woman and twelve months have passed and yet she is confined to her house. If you want her activities to be stopped, you can do it in many other ways.[40]

While she was under house arrest, she was adopted as 'Prisoner of Conscience' by the Amnesty International Group in England for her non-violent espousal of the cause of human rights.[41] The following years she was enrolled as an honorary member of Amnesty International.[42]

Though the Indo-Pak war of 1965 came to an end, it took a long time for the two countries to come to terms with one another. Through the mediation of Soviet Russia, the Prime Minister of India, Lal Bahadur Shastri and the President of Pakistan, Mohammed Ayub Khan agreed to meet at Tashkent in the Soviet Union to resolve their differences. An agreement was arrived at by the two Heads of State of 10 January 1966 which was known as the Tashkent Declaration. Soon after the signing of the Declaration, came the shocking news of Lal Bahadur Shastri's death in Tashkent due to heart failure. This caused a political change in India. Indira Gandhi was sworn in as Prime Minister of India on 19 January 1966. She removed the restriction of the movements imposed on Mridula. The orders of detention served on her were cancelled on 29 March 1967 enabling her to come out of her internment after a period of about one year and five months. Among those who congratulated her when the internment order was withdrawn were Indulal Yagnik, Syed Badruddya, Yunus Saleem, Mohammad Yusuf, Maulvi Abdul Gani Dar, Ebrahim Sulaiman and Yogendra Jha. S.M. Joshi, the socialist leader sent her a telegram. 'Your freedom gives encouragement to all freedom loving persons.' Sapru and Humayun Kabir in their message said, 'Hearty congratulations. Your correct, brave, patriotic stand was justified.'[43] On 31 March 1967 she wrote to Sheikh Abdullah from Ahmedabad as under:

Last week due to postal holidays, my posts got dislocated. Perhaps, by now you would have received my telegram sent to you on 29th intimating about my release. On the 29th evening at about half-past six, a police officer served me with an order of the Government of India, cancelling all the previous orders. Thus, my internment came to an end. Since then, I am calling on friends and going to Delhi. While I am here, I thought I better get inoculated against small-pox, cholera and typhoid, as sooner or later I will have to go to Bihar to meet Vinobaji and J.P.

Soon after the detention orders were withdrawn she went to Katni in Madhya Pradesh, met Vinoba and travelled with him through the villages of Madhya Pradesh and Uttar Pradesh trying to persuade him to take an active interest in Kashmir.

On her return to New Delhi, Mridula once again started her campaign for Sheikh Abdullah's release and pursued her efforts to secure the support of Members of Parliament. She was able to win over at first fifty-one and later 125 members who signed an appeal for Abdullah's release. The appeal ran as under:

As members of Parliament and representatives of the Indian people, we consider it our duty to ask for the immediate release of Sheikh Abdullah. The case instituted against him in 1958 was subsequently withdrawn and he has been almost continuously in detention since long. Consideration of democracy and civil liberty demand that Sheikh Abdullah should either be released forthwith or brought to trial on a specific charge without delay.

Amongst the signatories were Professor N.G. Ranga (Swatantra, Andhra Pradesh), Professor Humayun Kabir (Bangla Congress, West Bengal), K. Anbazhagan (D.M.K., T.N.), A.K. Gopalan (Communist (M), Kerala), Bhupesh Gupta (Communist, West Bengal), Indulal Yagnik (M.G.J.P., Gujrat), P.N. Sapru (Congress, U.P.), Anand Narain Mulla (Independent, U.P.), H.N. Mukerjee (Communist), Nirlep Kaur (Akali Dal, Sant Group, Punjab), G.S. Singh (Independent, U.P.), M. Ruthnaswamy (Swatantra, Madras).

Again the same appeal was signed by seventy four more Members of Parliament on 3 July, 1967. Among them were, S.A. Dange (Communist, Maharashtra), K.P. Singh Deo (Swatantra, Orissa), D.R. Parmar (Swatantra, Gujarat), Bhupindra Singh (Akali Dal, Punjab), Kizhedath Damodaran (Communist, Kerala), and others.

A Solution

The government finally released the Sheikh, as he was suffering from diabetes and other ailments. He was brought to Delhi in July 1967 for treatment at the All India Institute of Medical Sciences and all restrictions on his movements were lifted. He issued the following statement to the press:

To the large number of Members of Parliament from both Houses and of all parties who signed a representation for my release I owe a debt of gratitude. I am glad that the Government of India decided in response to that representation to withdraw the restrictive order served on me in May 1965. I must refer to the part played by several eminent friends outside Parliament who persistently worked for the same objective. While I cannot mention all, I must single out Rajaji and Shri Jayaprakash Narayan, Shri Shiva Rao and Shri J.J. Singh for their hearty support. I can assure you that notwithstanding the many years of my incarceration, I have no rancour in my heart. Personal liberty, after nearly fourteen years of unjust detention weighs far less with me than the release of forces that will advance the cause of friendship between India and Pakistan.... As long as I am alive, my life will be dedicated to the cause of Indo-Pakistan friendship and of inter-racial and interreligious unity. Regarding Kashmir I would say that in such friendship lies the hope of my State's peace and prosperity.

As restrictions on the movements of Sheikh Abdullah had now been removed, he visited Kashmir off and on. By this time, he seemed to have somewhat mellowed. While at Srinagar, he discussed with his close friends and well-wishers the line of action which should be adopted by him in respect of the Kashmir problem. The appeal signed by 125 Members of Parliament for his release from detention had boosted his morale and made him realize that there was a substantial group of politicians in India who still regarded him as their friend. He, therefore, began to think afresh about an amicable solution to the Kashmir problem and thought of holding a convention of the people of Jammu and Kashmir in order to find out the viewpoints of intellectuals and representatives of various political groups of the state. The convention was held at Srinagar between 10th and 15th October, 1968. He specially requested Mridula to attend as an observer. Jayaprakash Narayan was requested to inaugurate the convention whose main objective was to arrive at a lasting, honourable and equitable solution to the Kashmir problem. All participants were given an opportunity of expressing their views dispassionately and objectively. The Praja Socialist Party leader, Balraj Puri, was of the opinion that the real problem was between the State of Jammu and Kashmir and India. Pakistan had no locus standi in the matter. Sheikh Abdullah did not agree with that point of view. However, before the conclusion of the convention, it was decided to hold another convention some time in the near future. The proceedings of the convention were despatched to Vinoba Bhave, at the instance of Mridula, to acquaint him with the stand taken by different participants.

The second plenary session of the Jammu and Kashmir States People's Convention was held in June 1970 and Mridula was again requested to attend the Convention as an observer. Thus, Sheikh Abdullah kept his political activities alive but this time in a manner different from his previous method where he used to directly address the masses of the Kashmir Valley to bring them round to his view-point. In New Delhi too he had started his political activities by en-listing the support of J.J. Singh, the founder member of the Indo-Pak Amity and Friendship Group. In this way J.J. Singh had begun to take an active interest in Kashmir politics in the company of Jayaprakash Narayan.

On 9 January 1971, just before elections were to take place in Kashmir, Indira Gandhi, who was Prime Minister, served an order prohibiting Abdullah, Beg and G.M. Shah, the Sheikh's son-in-law, from entering Jammu and Kashmir. It was suspected by some that this was because there was a possibility of the Sheikh sweeping the polls. Three days later, the Plebiscite Front was declared illegal. The Government of India was backing the existing government of Jammu and Kashmir State. Mridula disapproved of all this. Mubarak Shah was arrested because he was a candidate from Baramulla con-stituency. The repression and arrests upset Mridula. But unlike on earlier occasions, this time she was not alone. Jayaprakash Narayan had now started playing a leading role and was having direct talks with Abdullah. A few of their nominees who did not belong to any political party stood for elections as Independent candidates. One such important candidate was Shameen Ahmed Shammem who contested the Srinagar parliamentary constituency against Bakshi Ghulam Mohammed. He was supported by the fol-lowers of Sheikh Abdullah. Shameen Ahmed Shammem won the elec-tion by a comfortable margin.

In 1971, general elections were also held in Pakistan. The Awami League of East Pakistan secured a majority of seats and claimed that the post of Prime Minister of Pakistan should go to their leader, Sheikh Mujeeb-ur-Rahman. This was unacceptable to President Yahya Khan and the leaders of all political parties of West Pakistan. Thus, acute tension began to develop between the leaders of West Pakistan and East Pakistan. A Non-Co-operation Movement was launched by the masses of East Pakistan against the Central Government of Pakis-tan. President Yahya Khan ordered his army to suppress the rebellion in East Pakistan. This caused resentment throughout East Pakistan. Repression of the highest degree was let loose on the civil population of East Pakistan by the Pakistan army. All the leaders of the Awami League including Sheikh Mujeeb-ur-Rahman were arrested and im-prisoned. Sheikh Mujeeb was taken to Lahore and put behind bars. People in East Pakistan began to flee from their homes en-masse to escape the brutality of the Pakistan sodiers. A massive number crossed the border into India. India had no alternative but to open its border on the eastern side to provide shelter to unarmed and innocent citizens of East Pakistan. Convinced that India was abetting the rebellion in East Pakistan, President Yahya Khan declared war against India. In that war Pakistan suffered a crushing defeat. Its forces had to surrender to the Indian army at Dacca on 16th December, 1971. This led to the formation of Bangladesh as an independent nation. President Yahya Khan had to step down and Zulfikar Ali Bhutto assumed power in Pakistan. About ninety three thousand military and civil personnel of Pakistan were made prisoners in Dacca. The wives and relatives of the imprisoned army officers raised a hue and cry for their release from the custody of the Indian army. Prime Minister Zulfikar Ali Bhutto could not ignore their pleas and had to rush to India to enter into an agreement with Prime Minister Indira Gandhi which is known as the Shimla Agreement. In accordance with that Agreement, all Pakistani prisoners were released and all disputes between Pakistan and India, including Kashmir, were to be resolved bilaterally. Sheikh Abdullah too realized that a truncated Pakistan no longer remained a power to reckon with. He, therefore, thought it advisable to come to some sort of an agreement with the Government of India.

Having taken his friends and well-wishers into confidence, he entered into negotiations with Prime Minister Indira Gandhi directly. Her Principal Secretary, P.N. Haksar fixed the date for the meeting and the talks were apparently held in a very cordial atmosphere. Thereafter, several meetings were held between them. Finally, it was decided that his nominee, Mirza Mohammed Afzal Beg and her representative in the Ministry of External Affairs, G. Parthasarathi should continue further negotiations. This was welcomed by the Indian press as well as the people of India. The political climate in the country had radically changed. Mridula was pleased that negotiations were under way between Sheikh Abdullah and the Government of India. In fact, this had been her objective from the very beginning.

She remained in constant touch with Sheikh Abdullah during the entire period of negotiations. Besides, she kept him informed about the favourable change in the mood of the people of India as would appear from her letter, dated 30 August 1973.

Here is an extract from the letter, dated the 11th August that I have received from Shri Porus Mehta for your information: 'I was in Poona when Sheikh

Abdullah came to Bombay. He got a "royal" welcome, staying in the Raj Bhavan. He is behaving like an elder statesman and yet not losing his old fire as the Lion of Kashmir. I do hope in his life time, Kashmir's future is settled and he is able to do something constructive for his native land.

That Mridula continued to be in close touch with Sheikh Abdullah is evident from her letter to him dated 16 August 1974;

From newspapers it appears that the recent conference with Jammu and Ladakh representatives went off satisfactorily. I hope this is true. It also appears that we are nearing solution of the long standing dispute. All the best wishes for the success ... I also read in the newspapers that you are thinking of going to Ladakh and your programme is being finalised. Is it not possible to fly to Leh and then motor down to Kargil and then fly again from Kargil to Srinagar? In army plane this may be possible. This is to avoid the height of Zojila pass. The other areas are also at higher altitudes, but the journey would not be strenuous. It can be undertaken provided you agree to spend two quiet restful days at Leh before you start working. It will only mean lengthening the stay for two or three days.

Mridula was always concerned about the health and well-being of others. At the slightest hint of illness, she would advise friends to consult doctors and get medical treatment. For a large number of persons, she had arranged treatment at the All India Institute of Medical Sciences, New Delhi. But for her own health, she was unmindful. During the month of October 1974 there was a sudden deterioration in her health. She breathed her last on 27 October 1974 and did not witness the 'Kashmir Accord' signed by Sheikh Abdullah and Indira Gandhi on 24 February 1975 for which she had struggled for so long. In accordance with the terms of the 'Accord', Sheikh Abdullah accepted the accession of Kashmir to India as complete and final. In return, he was once again appointed as the prime minister of Jammu and Kashmir State.

CHAPTER TEN

THE CRUSADER

Mridula Sarabhai started her public life at the age of twenty and carried on almost till her last breath. She fought for many causes—the country's freedom, Hindu-Muslim unity, rehabilitation of refugees and recovery of abducted women. From a young age, she was deeply concerned about women and the injustices they suffered both within the family and outside.

Like so many women of her generation, she was drawn to and inspired by Gandhi's thought and activities. The Sarabhai family frequently visited Sabarmati Ashram. Gandhi stayed at the Retreat on a few occasions and Mridula's aunt, Anasuya, was a close follower of Gandhi. Gandhi played a crucial role in drawing women into the freedom movement. He evolved a programme of struggle which included all sections and most significantly women: 'It was a stirring spectacle. That of thousands of women, who for centuries were chained to the narrow domestic life and whom an authoritarian social system had assigned the position of helots at home, stepping out into the streets and marching with their fellow male patriots in illegal protest demonstrations'.[1] No woman had been included by Gandhi in his chosen seventy one marchers to Dandi and this aroused considerable resentment among nationalist women. Among those who protested were Kamaladevi Chattopadhyaya, Khurshed Naoroji and Mridula. The Congress Committee withdrew its ban on women's participation and on the last day of the Salt march, Sarojini Naidu joined it at Dandi, and was the first woman to be arrested. Mridula was one of the thousands of women all over the country who actively participated in the Salt *satyagraha* and the Quit India movement and went to jail several times. She realized, however, that mere participation in the freedom struggle was not going to liberate women. In the nationalist movement, not only in India but elsewhere also, class, caste and gender issues were deliberately blurred in order to draw people into a countrywide alliance against a common enemy—foreign rule. In the first phase of the women's movement, nowhere were patriarchal values challenged. In India, the movement for women's emancipation was initiated by men and confined mainly to upper class/caste women and never, unlike in the West, did it arouse male hostility. Women leaders like Sarojini Naidu while pleading for political and other rights for women, repeatedly in public stressed women's primary role as good mothers and wives.

Evaluation of Gandhian ideology, has, on the one hand emphasized his contribution in raising the status of women. Gandhi saw women as autonomous, independent people, and also as an important base for his movement. His concept of *ahimsa* (non-violence), constructive activity and *satyagraha* were suited to what he defined as the characteristic qualities of women—love, non-violence and self-sacrifice. 'If non-violence is the law of our being, the future is with women' he asserted.[2] According to Vina Mazumdar, Gandhi's greatest contribution 'lay in his revolutionary approach to women in society'. He respected their 'personal dignity' without 'belittling their roles as mothers and wives'.[3] Madhu Kishwar holds that Gandhi saw women 'not as objects of reform and humanitarianism, but as self conscious arbiters of their own destiny'.[4] For most of the women who were active in the 1930s and 1940s nationalism had a beneficial influence on feminism and the freedom struggle, and according to them, liberated women.[5]

More recently, Gandhi's role as an emancipator of women has been questioned. Gandhi saw women as long suffering vessels of sacrifice. In this, he was not entirely original. Suffering and sacrifice were two qualities associated with the ideal Indian woman by Bankim Chandra Chatterji, Aurobindo and a number of early nationalists. But there was a difference. While reformers saw sacrifice for women as ritually enforced and hence to be deplored, and revivalists glorified sacrifices enjoined by tradition, Gandhi defined it as a special quality of Indian womanhood, based on women as mothers. Mothers bore the pain of pregnancy and sacrificed for their children, hence they could bear suffering and sacrifice for the country.

Gandhi did not challenge patriarchy and did not wish to disrupt the balance at home. According to him, men and women were biologically different and had different roles to play. The roles were complementary and each was equally important. The man's role was as bread winner, the woman's as wife and mother. In this, Gandhi was in the same tradition as Indian social reformers from Ram Mohan Roy onwards who wanted to improve the position of women but did not visualize any radical transformation in women's role within the home or outside. But while earlier reformers had implied that Indian women were ignorant, backward and oppressed, Gandhi emphasized the virtues of Indian women and said that Indian men as well as western women had a lot to learn from them.

While Gandhi found for 'women a new dignity in public life, a new confidence, a new self-view',[6] the role he assigned to women in the freedom struggle was mainly supportive and auxiliary, limited to spinning, picketing, distributing khadi, attending meetings and marching in processions.

Like a Gandhian, Mridula always wore *khadi*, no jewellery, no make up. Gandhi emphasized simplicity in dress and appearance and told women that they must no longer be 'dolls and objects of indulgence' but rather 'crusaders in common service' with their husbands. Women were not, according to him, sex objects. For those married men and women in his Ashram who wanted to totally dedicate themselves to the service of the country, Gandhi preached *brahmacharya* (sexless) lives and the unmarried, he asked not to marry or fall in love. Mridula's remaining single was partly Gandhi's influence and partly her fear that marriage was a bondage which would deprive her of her freedom. Mridula conformed to many of Gandhi's views but was much more radical in her perceptions regarding the role and status of women. She did not want women to be mere followers but decision-makers and this brought her into conflict with many nationalist leaders including Gandhi. She wanted women to be strong, independent and self-confident.

Apart from Gandhi, she was also influenced by Nehru whose perception of women was much more modern and progressive than Gandhi's. Mridula, while remaining close to both, charted out her own radical path for women.

In her last years, Mridula felt more and more lonely and isolated. After independence, she had worked for many causes—for the Kasturba Gandhi National Memorial Trust, for rehabilitation of refugees and recovery of abducted women. The last twenty years of her life she devoted to the Kashmir question and because of her anti-government of India stance, she lost many friends and colleagues. She displayed tremendous moral courage throughout but chose to trod a rough and lonely path. She confessed to Vinoba Bhave that things had become so confused that she was not clear as to what she could do.

... I think I sent you copies of letters I wrote to J.P. some time in July-August last year. He had called me to say that as my work in Kashmir was over and I could not be effective any more, I had better take up some other work. I had replied that in a way what he was saying is true. In the political world the work has been taken over by him and others so my role is only to 'stand and watch', yet I feel my role is not over. There is lot of constructive work to be done in Kashmir and I would like to do that. He wondered if Government would allow me to do this. I said I have to get my right to work in place of my choice. I did not think Government would come in the way. So I would like to do this as well as I would be quite ready to do any other activity with Delhi being the Headquarter. In writing I offered to work for any of the following: (1) Indo-Pak goodwill, (2) communal harmony, (3) creating congenial atmosphere in India for solution of knotty problems as Kashmir, Nagas, etc., for example, removal of tension in the hill areas, in other words be an active worker of their Shanti Sena's in border area, (4) lobbying in Parliament for Sarvodaya Programme, (5) educating the voter, etc. I have not got the copy of this letter with me at present but I remember having mentioned all this. It is now about 8 months—I have not been asked to work in any of these or any other activity.

... About a week ago Radhakrishnaji, Manmohanji and Narayanji spent a night with me here. I told them of my growing feeling that I am being tolerated in the Sarva Seva Sangh because of Vinobaji.... The Sarva Seva Sangh also is not serious about work in Kashmir a very delicate sensitive area in India: This is our misfortune.... I was able to convince them that even though political matters are being handled by J.P. yet there is much to be done by Sarva Seva Sangh, such as (1) creating congenial atmosphere for implementation of policies decided by J.P. and Sheikh sahib, (2) taking human problems and work for their solution, for example, refugee problem, border area problem, communal problem, students problem, etc....

... Working in Kashmir is not easy if one wants to do this on the principle of self help and lokshakti. The 15 years rule through corruption has resulted in complete degeneration of character. Then again the rule of repression and persecution has marred the spirit of initiative and urge to do anything constructive. People and leaders suffer from all this; acute cold war adds to problems. For me these fifteen years amongst other things have been a training in patience. It is being put to the maximum test. It has been suggested to me that I should give up this area and take up any other place. I feel this is a wrong policy. One should work where one has goodwill also. This I have in Kashmir as well as in Ahmedabad: I am a believer in workers specializing in areas or states. Then only concentrated work can be done. So why should I be asked to leave Kashmir....

... As to myself. For the first time in life I have the chance to decide what I should do. Until now I was invited and associated with such movements that evolved from one to another leading to bearing of responsibilities that had to be borne. Just now I am free of special responsibility that leads to the question what next? How much new responsibilities should one take? But a basic question that faces me is the assurance I gave to Bapu in Bihar. You remember I have often told you about this. He called his last mission as an experiment to find out an instrument for individuals and natives of free countries to resist internal forces of disruption of democracy. I gave him an assurance that I will work for this. The work done in Kashmir or for Kashmir was as a result of this assurance—now what? The whole country is facing a great challenge. I consider Bhoomidan movement a basic approach towards solution of Gandhiji's mission. But as I don't find myself up to it and joining you in your Bihar experiment and as Kashmir problem yet threatens to engulf the subcontinent, I am in search of a way as to what should be done. Perhaps the present position of 'stand and wait' may also be an activity towards that end. Giving priority to solution of political problem of Kashmir has become such a complex one that to get active cooperation for constructive work is difficult. Then again as I have mentioned, lack of fearless constructive organisations in Kashmir is a great handicap. So I am reconciling myself to the situation that Bapu and his colleagues faced in 1920 or so....

Earlier in the same letter Mridula recalls:

... I well remember your remark when we last met. You had suggested that I should at least meet you every fortnight or at least once a month. It is now about a year since we have met. This has often pricked my consciousness. The breach looks like 'disobedience', 'selfishness' and can be taken as ingratitude on my part: It is not 'giving you up' as some may say. The truth is that circumstances have become so confused and complex that I am not clear as to what I should tell you when I see you. It is true that now and then on personal level I have faced 'storms'—working out anything has been very difficult. During the last 18 years one faced obstructions from opponents. Now is added from within. When it comes to personal matters I am so used to facing it on my own that it does not become a subject for consultations. This is my habit. All throughout my life, I do not remember having gone to any one, even to my parents with my problems. Problems of work and activities are for consultations no doubt, and that took me to you also. Now everything has got mixed up—so until something is clear—until I see light—is it fair to take your time?—this consideration only has kept me back....[7]

Mridula was now unable to shoulder new responsibilities on her own, single handedly, as in the past and there was hardly an organization which was not apprehensive about associating with her. At fifty eight, her health was failing. Despite her brisk walk, militant stance, and indefatigable energy. Mridula was not physically a very strong person. She was subject to frequent illness, the years spent in jail both in the pre-Independence and post-Independence period and decades of roughing out doing field work further ruined her health. In 1963–4, she complained of gastrointestinal upsets and tests at the All India Institute of Medical Sciences revealed that she had amoebic cysts and she was treated for chronic amoebiasis. The following year, she complained of blurred vision and frequent headaches and it was diagnosed as high blood pressure. Between 1965 and 1968, she developed severe osteo-artheritis in both knees and also spondylitis but she never made any fuss over her ailments, in fact she underplayed them as much as possible and never informed her family of her frequent illness. Since 1971, her sister, Bharati's daughter, Anarkali who had some psychological problems, was staying with her in Delhi doing post-graduate studies at the Delhi School of Social Work. Guiding Anar took a great deal of Mridula's attention and energy.

In early October 1974 Mridula had a fall and with difficulty managed to get up and reach home. She felt a girdle of pain around the diaphragmatic area and in twenty four hours the pain intensified accompanied by high fever. X-rays did not reveal any fracture but showed evidence of pleurisy at the base of the right lung. On 18 October she was admitted to AIIMS and ten days later the liver became palpable and X-ray of the chest for the first time revealed an opacity which was diagnosed as an embolism, probably arising out of a malignancy in the liver. It was decided to get radioactive isotope from the Army Medical Centre and inject it to trace the source of the malignancy in the liver. The material was brought but by then, her general condition had deteriorated. Her brother, Gautam, who had by then arrived in New Delhi, after discussion with the doctors decided that since the test was being done only for an academic diagnostic purpose and was not going to benefit the patient and was likely to be painful, need not be done. He advised that the radio isotope should be used for some other patient to whom it could prove beneficial. The Director of AIIMS entirely agreed with this decision.

Mridula had been in a critical condition since 24 October. Her condition showed slight improvement on the night of October 25th but soon it took a serious turn. She was too weak to speak. With a faint smile on her lips, she looked at her mother, brothers, sisters, nephews, nieces and others who had gathered round her and fell into a coma. Her mother was persuaded to go home and rest for a while but was soon recalled to the hospital. Saraladevi was wheel-chaired to Mridula's bedside and as soon as she put her hand on Mridula's forehead, Mridula breathed her last at dawn on 26 October 1974.

According to the doctors and nurses who attended on her, she remained cool and patiently endured the pain without any fuss.

Her body was brought back to her Chanakyapuri residence, laid on the floor covered with a stupa of flowers, the face serene and calm:

Life's fitful fever had ended
Leaving not a rack behind.

The Prime Minister, Indira Gandhi arrived early in the morning and stood by Saraladevi who was looking stunned and woe racked, frozen as it were in a vigil of grief. Two months later, she wrote to Indira Gandhi: 'The shock was like a thunderbolt'.[8] This was the second time that she had to face the loss of one of her children. Sheikh Abdullah had flown in from Srinagar and was at Mridula's bedside when she died. It was ironical that on either side of her body stood Indira Gandhi and Sheikh Abdullah, the two persons she had wanted to bring together. Both of them walked out together.

Mridula's body was cremated at the electric crematorium after prayers. Among those present were Indira Gandhi, the Vice-President B.D. Jatti, the Petroleum and Chemicals Minister K.D. Malaviya, the Minister without portfolio Uma Shanker Dikshit, Sheikh Abdullah, Aruna Asaf Ali, members of her family and a large number of friends, colleagues and admirers. One of those present remarked that he had never seen a more secular gathering at a crematorium than on this occasion. More than half of those present were Muslims.[9]

Thus ended a stormy, often controversial career stretching over a span of forty years marked always by exemplary courage and a deep sense of compassion.

Even during her severe pain and illness, she could not forget the Kashmir problem. Unfortunately, the Kashmir accord was reached within a month of her death. Ironically, Indira Gandhi and Sheikh Abdullah travelled on the same plane to Srinagar on 30th October when Sheikh was carrying Mridula's ashes in an urn and they discussed the Kashmir issue.

Dr Farooq Abdullah recalls that when he went to see Mridula a few days before her death, she spoke not of her illness, but how much his parents had endured and suffered and that they, the children, must not cause them any further problems.

'Mridulaben was extremely feeble, could hardly talk and was whispering but her thoughts were about others, not about her own approaching end. This was typical of her, never thinking of herself.'[10]

Condolence letters and telegrams poured in from all over India and abroad. Indira Gandhi who had known her for over four decades wrote to Saraladevi—'Mridulaben was an extraordinary person and unobtrusively influenced many trends in India. She assumed a stern exterior but all who knew her also witnessed the strength of her friendship and the warmth of her sympathy.'[11]

Mridula was a rare human being—totally sincere, straightforward, self-effacing and simple, remarkably brave and strong, but at the same time gentle and compassionate, with a rare courage of conviction, spirit of public service, sacrifice and integrity.

Apa Saheb Pant wrote to Bharati:

I met Mridula first in 1940 at Vrindawan in Bihar. Immediately I came under her spell which has lasted these 34 years and would ever be with me.

And what a 'spell' it is! Character, dedication, loyalty, truthfulness, utter courage, and compassion—oh! how 'compassionate' and actively so she was. For her suffering meant direct immediate action. She was not only a true worker but a total worker and a real Yogini. Others, like me, talked of, philosophized, about revolution, service, action (direct or indirect), Mridula was it, all the time.

I have had intimate contact with her under all the most difficult, dangerous and distressing circumstances. I know the period when she was fighting 'Establishment' and was declared 'Mad' to be put into an asylum!! How she fought corruption, bigotry, communalism. She was a true Hindu—Muslim—Parsee—Sikh—Jain—Christian—a real human being.

And what an organiser!! During the 1946, 1947 and 1948 riots and abductions what she did and how she organised it is a legend in itself.

Well, Mridula is no longer there. Her inspiration will always, ... carry the perfumed memory of Mridula with them. As long as her type are born in India, India will be glorious.'[12]

MRIDULA SARABHAI

A Biographical Sketch

Permanent Address: 'The Retreat', Shahibag, Ahmedabad-380 004

Temporary Address: 31, Rajdoot Marg, Chanakyapuri, New Delhi-21

Born at Ahmedabad on May 6, 1911

Schooling done at 'The Retreat' in the school conducted by her parents specially for their children. Passed Vinit (equivalent to matriculation) in 1928 and joined Gujarat Vidyapeeth (the National Mahatma College) started by Gandhi. Passed the first year Snatak Course. Discontinued college studies in 1930 to join the national freedom movement. Continued her self-education in jails and outside and during her apprenticeship with her parents, Shri Ambalal and Smt. Saraladevi Sarabhai, Jawaharlal Nehru, Mahatma Gandhi and a few others.

Took part in agitations during the Freedom Movement in Ahmedabad, Gujarat, Saurashtra and later on throughout India from 1920 to 1947.

1920–25: Worked in the Bal Sena of the Congress and the Shishu Vibhag of Akhil Bharat Charkha Sangh.

1925–30: Joined the youth movement, under the leadership of Jawaharlal Nehru. In 1928 attended the Youth Conference at Rajkot under Jawaharlal's presidentship. Mridula was one of the organizers.

1930–46: Was a Congress Seva Dal volunteer in the Freedom Movement, participating in the Salt Satyagraha, organizing the boycott of foreign cloth and British goods; provided legal defence and did relief work for political workers and their families through organizations like Jyoti

Sangh; specialized in the constructive programmes of Qaumi Ekta (communal harmony), Stree Jagriti, and in journalism.

1934–7:   Was a delegate to the Indian National Congress; was an AICC Member for some time during this period. In 1935 came into open conflict with Congress leaders of Gujarat Provincial Congress Committee on the question of having voted in favour of Jawaharlal Nehru's resolution for non-participation in Parliamentary programme. This was at the Lucknow Congress session in 1934. The Gujarat Congress leaders wanted to take disciplinary action on the grounds that Mridula had broken the age-old traditions of the Gujarat Provincial Congress Committee, AICC and Delegates to Congress sessions to vote after watching how the leader voted. Mridula protested against this convention. She said that the procedure of previous discussion and mutual consultation with delegates had not been followed. They demanded her resignation from the AICC and as a delegate on the grounds that she had been allowed to get these positions on their having included her in their panels. So she was duty-bound to resign. She agreed to resign but made it clear that she would contest the election for delegates as an independent. Gandhiji intervened and the matter was dropped for the time being. During the next election Mridula contested as an independent candidate. The Congress voters rose to the challenge. They flouted the severe pressures brought upon them by the Congress bosses of the Gujarat Provincial Committee and the Majdoor-Mahajan Sangh (Labour Union) and voted for Mridula. She got the highest number of votes and was the first ever competing with the delegates elected from the official list, which included many stalwarts. However, she could not succeed in the AICC because of an official blockade. After that, her troubles in the Congress organization started. A number of times her primary membership forms were 'lost' in the local Congress Committee office. Restoration of this was essential, otherwise it would amount to a break in the continuity of membership leading to disqualification and therefore debarring her from becoming an 'active member' and taking away the right to vote and stand for election. Her membership often used to be restored only at the intervention of the President of the All-India Congress Committee or the All-India Working Committee.

After the delegates voted for her, the official block gave up resistance at the final voting stage. By then there were two clear groups amongst the Congress workers in Gujarat. One was the Sardarites or Patelites and the other Nehruites. Most of the Nehruites later became Socialists, Communists, etc., while Mridula herself remained a Nehruite.

From that time onwards Mridula could get into the AICC only when there were a sufficient number of 'opposition' members (as they were named) amongst the delegates.

From this experience and as a result of the treatment meted out to her and to some others within the Congress, it was shown that an anti-Nehru group was functioning. In those years this group had come to be known as the 'syndicate'. Mridula had doggedly resisted them and so became a special target all through until 1958 when she was expelled from the membership of the Congress on the ground of having violated Congress discipline!

While this persecution within the Congress organization continued, it must be said to the credit of the Gujarat Congress leaders and Mridula that their personal relations remained so good that most of the elder Congress leaders supported her and encouraged her in developing women's movements within and outside the Provincial Congress Committee, gave her important positions within the Congress as an in-charge of the Women's Department, Seva Dal, etc. Morarji Desai gave her an opportunity as an apprentice under him in various activities. Thus, though out of favour within the Congress organization, she was able to get acquainted with the whole of Gujarat and accepted as a trusted colleague by all parties.

1937:   Participated in the Rajkot Satyagraha as a Praja Mandal volunteer and after Tripuri Congress resolution worked closely with the J & K National Conference as an associate worker upto 9th August, 1953.

1937-8: Became a member, Municipal School Board, Ahmedabad.

1937-9: Was a member, Bombay Government Prohibition Board and the Development Board; Secretary, Sub-Committee for Women's Role in the Planned Economy of the National Planning Committee of the Congress under the Chairmanship of Jawaharlal Nehru. During all this period she was in-charge of the Women's Section, Congress Seva Dal, Gujarat, also in-charge of the Women Volunteers at the Haripura Congress and on the AICC committees for these activities.

1941-2: She was acting Secretary, City Congress Committee, Ahmedabad

1945-6: She became organizing Secretary, Kasturba Gandhi National Memorial Trust. Later on, she was a Shanti Sainik in Gandhiji's programme of national and emotional integration. In 1945, she was offered membership of the Bombay Legislative Assembly but under Gandhi's advice decided to remain out of the parliamentary programme and continued to work as a Shanti Sainik.

1946: In 1946, when Jawaharlal Nehru became the Congress President, he nominated Mridula as a member of the Congress Working Committee and appointed her as one of the General Secretaries of the Congress. This lasted until he become the Prime Minister of India in the Interim Government and at the Meerut Congress a new Congress President took over.

Mahatma Gandhi and the national leaders under whom Mridula had her political apprenticeship had always emphasized that everyone should work with the entire nation's welfare in mind and be prepared to do one's best in the areas that needed workers the most. Gujarat has always had a good network of workers, so Mridula took the opportunity to work outside Gujarat. This and her efforts to specialize in Gandhiji's Shanti Sena movement in 1946 when he was leaving Noakhali, resulted in her doing peace work in Bihar, East Punjab, Delhi, Uttar Pradesh, West Bengal, Jammu and Kashmir and in both wings of Pakistan.

During the communal disturbances, she specialized in organizing active resistance to them. Participated as a Shanti Sainik in bringing about normalcy during the following communal disturbances and skirmishes. Small skirmishes are not mentioned as they were many but the major ones were:

1941	:	Ahmedabad communal riots
1946	:	Garh Mukteshwar (Meerut)
1946–7	:	Bihar
1947–8	:	Punjab, both East and West
1948–50	:	Delhi
1950–51	:	Bengal, both East and West
1961	:	Aligarh
1969	:	Ahmedabad

1946–7:   In Bihar she worked with others in charge of the peace mission and the rehabilitation of uprooted minorities. From 19 August 1947 she was in the disturbed areas of East Punjab and West Pakistan, specially West Punjab. Started working as a peace worker in both Pakistan and India. Secured full cooperation in this mission from governments and people of both the dominions, resulting in development of the following activities.

Until Bapu's Shahadat (martydom) on 30 January 1948 she worked under his direct supervision. Later on both the governments created special agencies for these activities. So she was allowed and encouraged to function for the cause by both the governments. Worked under the direction and general supervision of Jawaharlal and Vinoba Bhave. Did moral, social and economic rehabilitation work as a member of the U.C.R.W. (United Council for Relief and Welfare) under the guidance of Lady Mounbatten when C. Rajagopalachari was Governor General of India and Dr Rajendra Prasad was President of the U.C.R.W.

As far as working with both Governments was concerned, she was under Shri Gopalaswami Aiyanagar and

Sardar Swaran Singh in the Government of India and in Pakistan under Mian Iftikharuddin, Raja Ghaznafar Ali Khan, Ghulam Mohammed Saheb, then Finance Minister and Khan Qurban Ali, IGP. As a common worker in India and Pakistan she helped their organisations in the recovery of abducted women and in the care of minorities.

1950–53: She was the Convenor, Friends of New Kashmir Committee, which organized Delhi citizens' resistance to the communal movement in Kashmir.

1953: She was Convenor, Shanti Dal, Delhi.

1948–53: She was in-charge of special rehabilitation schemes of the Jammu and Kashmir Government on behalf of the National Conference. Worked under the direction of Gopalaswami Ayyangar and Mirza Mohammed Afzal Beg and under the general guidance of Sheikh Mohammed Abdullah.

During this period the Congress leadership worked at cross purposes. One section inspired by Jawaharlal was more secular in its approach while the other, was inspired by pro-Hindu and anti-Pakistan sentiments. Through her activities, Mridula exposed and resisted these elements within the Congress and in the government.

After Sardar Patel's death, these elements' open resistance came to an end and they claimed that they supported Nehru. Their first targets were the original Nehruites like Sheikh Abdullah, Mridula and others. The uprooted and disruptive forces in and out of the country utilised this opportunity to undermine Nehru's efforts to build a secular outlook and approach in and out of the government. Kashmir became their first target.

9-8-53   During these tragic events in Kashmir she came into conflict with these disruptive and vested interests within the Congress and outside, and their challenge to Gandhi's last mission in which he wanted to establish a technique of how people should resist the internal forces of disintegration in government and in society. Therefore, as an individual resister she carried on the activities of a Shanti Saink in Jammu and Kashmir State and for the people of

Jammu and Kashmir in the rest of India.

It was during Jawaharlal Nehru's regime that many non-opportunist Nehruites were subjected to severe persecution, character-assassination and damage to their public image. To some extent these anti-Nehru, anti-democratic forces within the Congress and the government succeeded. Mridula had often been asked, 'How is it that Nehru took action against you when you were such a trusted worker and friend of his?' From all that had come to light, Mridula's conviction had deepened that many things were done in his name and that of the government by individuals in the government and in the party, whom Nehru trusted for 'truthful' reporting and implementation of his policies without at all suspecting their internally disruptive activities. He had been stabbed in the back. In later years he appears to have realized that he had been terribly let down. It is for this reason that the victimized Nehruites continued to hold him in great esteem and not responsible for their trials and tribulations.

Experience in Journalism

Worked as a reporter for the *Gujarat Samachar* in Ahmedabad and for the National Herald. For years she was on the Board of Directors of the publishing houses of both these papers. During the movement for independence she worked for the Congress Patrika and after 1955, to break through the press and information black-out on Kahmir, she issued circulars on the Kashmiri people's version of their trials and on their appeals for intervention. Occasionally, wrote articles for the press on topical subjects.

She concentrated on cultivating public opinion in favour of resolving the Kashmir dispute with the support of all concerned, specially for finding a solution to the internal conflict that had been going on in Kashmir since 9 August 1953. This led to her being involved in a political case as a co-conspirator by the Bakshi regime. This was helpful. It gave her a greater opportunity to place her case before the bar of public opinion.

Jails and Detentions

Due to the British Government's Policy in Gujarat of not arresting women unless forced to do so and the orders to her by those in charge of conducting the movement in those days not to participate in active *satyagraha* but to do organizational work, she was arrested only three times. Usually she was detained while working. Earlier in 1930 she participated in many demonstrations but was always released with others except once when she was detained for three weeks. Periods of detention and imprisonments were:

Pre-Independence

1930:   3 weeks, imprisoned.

1932:   6 months' detention.

1932–3:   Externed from the walled city of Ahmedabad and then detained. Released in Belgaum and placed under house detention. Joined the local Congress committee demonstrations on 13 April–National Week. Arrested and convicted for six months with a fine of Rs 300/-.

1937:   In Rajkot Satyagraha movement, she was detained for about 3 months.

1942–4:   Detained in Bombay on about 20 August. Unconditionally released in December 1944 on medical grounds.

All through these periods she was kept in Sabarmati Jail, Yervada Jail, Belgaum Jail and Arthur Road Jail, Bombay.

Post-Independence

1958–9   Detained under the Preventive Detention Act in Tihar Jail, Delhi.

1965-7:   House detention at Ahmedabad under the Defence of India Rules.

1953–74:   During this period, she came into conflict with the policy of the Government of India to remove Sheikh Mohammed Abdullah from the post of Prime Minister of Jammu and Kashmir State and detaining him in jail. She relentlessly campaigned for his release by influencing public opinion in his favour through the issue of her circular letters. When the trial of Sheikh Mohammed Abdullah started, she associated herself with the Legal Aid committee that was formed by her collegues. She firmly believed that Sheikh Mohammed Abdullah was the supreme leader of Jammu and Kashmir State and no solution of the Kashmir tangle could be found without his cooperation. The Government of India ultimately veered round to her viewpoint and released Sheikh Mohammed Abdullah unconditionally. Additionally, he was made the Chief Minister of Jammu and Kashmir State in February 1975 and restored to power with full honour and grace. Unfortunately, she could not witness that ceremony as she expired in October 1974 following a sudden illness.

1970–73:　She associated herself closely with Khan Abdul Ghaffar Khan, (the Frontier Gandhi), who visited India in 1969 in connection with the Gandhi Centenary Year and assisted him in the formation of a society which was named Insani Biradari (Human Brotherhood). she served the Insani Biradari with great enthusiasm and zeal.

1968–74:　Due to her untiring efforts, an Indian branch of Amnesty International was started in India in 1968. Mr M.C. Setalvad became its Chairman and she became one of its secretaries. She continued to serve in that organizaion till her untimely death in October 1974.

Appendix I

LIST OF PERSONS INTERVIEWED

1. Abdullah, Begum, widow of Sheikh Abdullah, Srinagar.
2. Abdullah, Farooq, son of Sheikh Abdullah, former Chief Minister, Jammu & Kashmir.
3. Bach, Puratan, educated at Gujarat Vidyapith. Freedom fighter and friend of Mridula Sarabhai.
4. Chattopadhyaya, Kamaladevi, 1903–88, fought for the country's freedom and was a champion of women's rights. She was one of the founders of the All India Women's Conference and its first General Secretary. She was founder President of the Indian Co-operative Union; active in the field of arts, crafts, theatre.
5. Desai, Devyani (Devibehn), social worker, freedom fighter, Vile Parle, Bombay.
6. Desai, Vanmala, daughter of Narahari Parikh, wife of M.V. Desai. She was brought up in Sabarmati Ashram. Took part in Quit India Movement. Author of several books. New Delhi.
7. Gosalia, Pramoda: born 24.11.1913. BA SNDT University of Bombay; sister of Sarladevi Sarabhai. Joined Student's Wing of Youth League of Ahmedabad in 1928. Participated in Salt Satyagraha. Active member of Jyoti Sangh since its foundation. Member AIWC and various women's and social welfare organizations, Bombay.
8. Hegishte, Hemlata, social worker; took part in freedom struggle; associated with Jyoti Sangh; colleague of Mridula Sarabhai.
9. Jayakar, Pupul, Cultural Adviser to Indira Gandhi, Prime Minister of India. Author of 'Earthen Drum', and biography of J. Krishnamurti.
10. Joshi, Subhadra, ex-Member of Parliament, formerly General Secretary., Delhi PCC.
11. Karan Singh, (Dr) Sadar-i-Riyasat, 1952–65; Governor, Jammu and Kashmir, 1965–7; Health Minister, Government of India. India's Ambassader to U.S.A.
12. Kartikeya Sarabhai, son of Vikram Sarabhai.
13. Kercood, Vayusutha, social worker, close associate of Mridula Sarabhai, Hyderabad.

14. Leena Mangaldas, daughter of Ambalal Sarabhai, sister of Mridula Sarabhai; educationist, founder of 'Shreyas', Ahmedabad. Author of *Akhand Divo.*
15. Medh, Parashar, Lawyer, Ahmedabad.
16. Medh Susmita, Writer, Author, Ahmedabad.
17. Mehta, Usha (Dr), former Professor of Political Science, Bombay University, took part in 1942 Quite India Movement; Director, Mani Bhavan, Bombay.
18. Pant, Apa Saheb, Raja of Auadh, formerly India's Ambassador to various countries including Egypt, Britain, etc. Poona.
19. Raja Rameshwar Rao, former MP Orient Longmans, Hyderabad.
20. Radhakrishna, Gandhi Peace Foundation, New Delhi.
21. Ranganathan, Surekha, daughter of Mrs Kercood, Hyderabad, Assistant Divisional Manager, MMTC.
22. Sarabhai, Gautam, brother of Mridula Sarabhai.
23. Sarabhai, Geeta, sister of Mridula Sarabhai.
24. Sarabhai, Gira,, sister of Mridula Sarabhai.
25. Sarabhai, Mrinalini, (nee Swaminathan) wife of Vikram Sarabhai.
26. Sarabhai, Anand, son of Suhrid Sarabhai, nephew of Mridula Sarabhai.
27. Sen, A.C., Secretary, Amnesty International, India.
28. Singh, Khushwant, journalist, writer.
29. Singh, Malti, wife of J.J. Singh; President, Amnesty International, India.
30. Teng, Muhammad Yusuf, Secretary, Cultural Academy, Srinagar.
31. Thakore, Harivadan, Congress worker freedom fighter. Worked with Mridula Sarabhai at Haripura Congress.
32. Thakore, Rohini (see Desai), daughter of Hitendra Desai, ex Chief Minister of Gujarat, friend and associate of Mridula Sarabhai.
33. Tyabji, Badruddin, ICS, India's Ambassador to Idonesai, Iran, West Germany and Japan. Special Secretary, Minister of External Affairs, Author.

The following were kind enough to respond to my letters.

G.M. Shah, son-in-law of Sheikh Abdullah, Minister in J&K Government at various times. Sadr-ud-din Mujaheed, Chairman, J&K Khadi and Village Industries Board, Vice-President, All India Freedom Fighters Organization; President, All India Freedom Fighters' Organization, J&K.

The following shared some of their impressions of Mridula Sarabhai with me.

The late Shri Umashankar Joshi, noted poet, author, President Sahitya Academy, Member, Rajya Sabha. Prof. Bimal Prasad, SIS JNU. New Delhi, Mrs Aruna Asaf Ali, Mrs Gulab Pandit, Dr Sushila Nayar, Prof. V.N. Dutta.

Appendix II

MAHATMA GANDHI'S LETTER TO MRIDULA SARABHAI

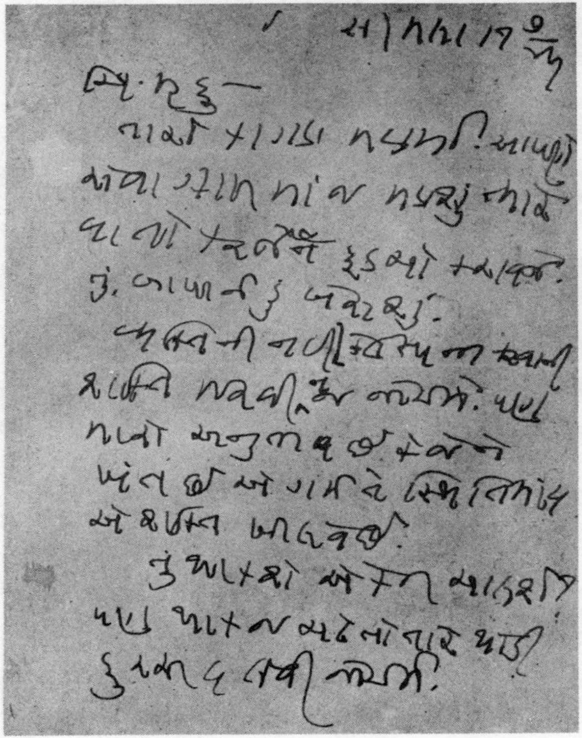

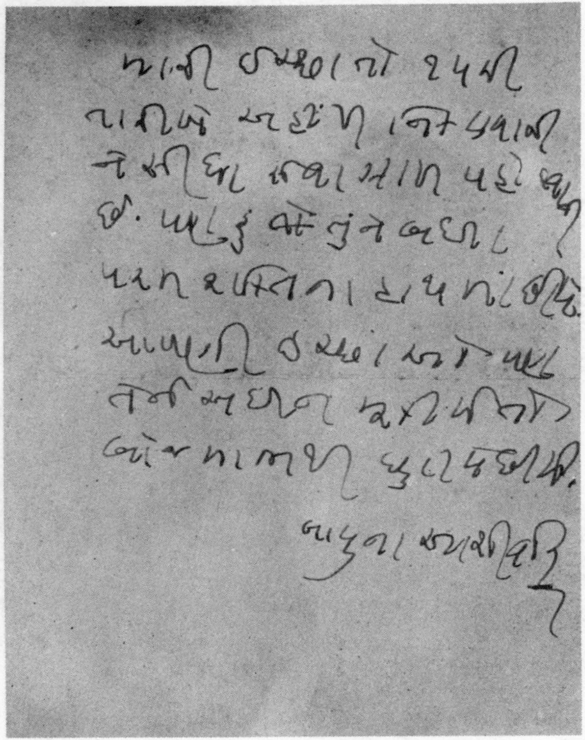

Appendix III

From Jawaharlal Nehru
Dihict Gail, Dehra Dun

To Mridula Saratai
The Retreat, Shahibagh
Ahmedabad.

June 4. 1941

My dear Mrida,

India told me yesterday that you wanted to have an interview with me. Apparently you are likely to come up to Lucknow after a week or so and you intend running up here then. I shall of course be happy to see you and I find that it will be possible to have an interview about the 15th or 16th of this month. So you can write or wire definitely as soon as you are sure of your programme.

You must have had a difficult time on account of the communal troubles in Ahmedabad. These are very distressing and must of course be faced. But there is no need to despair or grow faint-hearted about such matters. We have to harden ourselves and remember that the world is going through a bad phase.

Love

Yours affly
Jawaharlal

Appendix IV

Number of Women and Children recovered from Pakistan between 6.12.47–31.8.55

		6-12-1947 To 31-12-1949	1-1-1950 To 31-12-1950	1-1-1951 To 31-12-1951	1-1-1952 To 31-12-1952	1-1-1953 To 31-12-1953	1-1-1954 To 31-12-1954	1-1-1955 To 31-12-1955	Total
1.	Punjab (Pakistan)	4981	287	234	116	70	60	53	5801
2.	North West Frontier	413	23	8	16	25	3	5	493
3.	Baluchistan	10							10
4	Sind	30	12	9	5	16	4	8	84
5.	Native Indian States	579	1	10	4	2	4	1	601
6.	Jammu & Kashmir	259	548	482	333	211	89	121	2043
	Total	6272	871	743	474	324	160	188	9032

NOTE: The Kunja Camp in Gujarat had the largest number of recovered women and children—950, followed by 818 in the Sialkot Camp.

SOURCE: Appendix 1 in Kamlaben Patel. *Mul Sotan Ukhdelan (Uprooted)*. Ahmedabad: 1985.

Number of Muslim Women and Children recovered from India, 6-12-1947–31-8-1955

		6-12-1947 To 31-12-1949	1-1-1950 To 31-12-1950	1-1-1951 To 31-12-1951	1-1-1952 To 31-12-1952	1-1-1953 To 31-12-1953	1-1-1954 To 31-12-1954	1-1-1955 To 31-8-55	Total
1.	Punjab	8724	698	1240	571	984	491	213	12,931
2.	Delhi and Surrounding areas	62	2						64
3.	PEPSU (Faridkot, Jind, Kapurthala, Natha, Patiala & other states	3280	534	695	480	918	575	225	6,707
4.	Rajasthan (Alwar, Bharatpur, etc.)	275	17	15	3	10	5	10	335
5.	Jammu & Kashmir	211	162	24	108	128	43	25	701
6.	Total	12,552	1413	1974	1162	2040	114	473	20,728

SOURCE: Appendix 2 in Kamlaben Patel. *Mul Sotan Ukhdelan (Uprooted)*. Ahmedabad: 1985.

NOTES

Introduction

1. Arun Gandhi, Mridula Sarabhai *Illustrated Weekly of India,* November 1995.
2. Mridula Sarabhai, *Gujarat Samachar,* 3 November 1974.
3. Jawaharlal Nehru (henceforth J.N.), Prison Diary, *Selected Works of Jawaharlal Nehru* (henceforth *SWJN*), vol. 13, p.13, p. 178.
4. Ibid, p. 178 and Ibid. vol. 11, pp. 618–19.
5. Usha Mehta, Mridula Sarabhai, *News Services AIR,* nd.
6. Chhotubhai Bhatt, Mridula Sarabhai, *Jivan-Jhanki.* Kasturba Darshan, May 1975.
7. *Collected Works of Mahatma Gandhi* (henceforth *CWMG*), 14 January 1948, vol. XC, p. 424.
8. Mridula Sarabhai's (henceforth M.S.) talks on All India Radio, 15 June 1969.
9. J.N. to Bharati Sarabhai, 10 March 1938 *SWJN,* vol. 8, pp. 864–5.
10. J.N. to Indira Gandhi, 19 June 1941, *SWJN,* vol. 11, pp. 618–19.
11. Ibid. p. 618.
12. J.N. to Padmaja Naidu, 25 March 1938, *SWJN,* vol. 13, p. 647.
13. M.S. to J.N., 7 June 1959 (strictly personal).
14. J.N. speech in Lok Sabha, 9 December 1958.
15. M.S. to Aryanayakans, 9 June 1964.
16. Erik Erikson, *Gandhi's Truth,* New York, 1969, p. 22.
17. H.V.R. Iyengar, *Sunday Standard,* 27 September 1964.
18. Apa Pant to Bharati Sarabhai, 2 November 1974.
19. J.B. Kripalani to M.S., 30 August 1941.
20. J.B. Kripalani to M.S., March 1934.
21. J.B. Kripalani, Interview given at Sheth Chaturbhujdas's Bungalow, 19 June 1976, M.S. papers.
22. Iyengar, *Sunday Standard,* 27 September 1964.
23. Gail Minault, *The Extended Family: Women and Political Participation in India and Pakistan,* Delhi, 1981.

24.   A.D. D. Gordon, *Businessmen and Politics: Rising Nationalism and a Modernising Economy in Bombay, 1918–33,* New Delhi, 1978. Dwijendra Tripathi and Makrand Mehta, 'Class Character of the Gujarati Business Community' in D. Tripathi, ed., *Business Communities in India,* New Delhi, 1984.

Chapter One: Early Influences

1.   Kenneth Gillion, *Ahmedabad: A Study in Indian Urban History,* Berkeley, 1968, pp. 13–15 and 29; and *Ahmedabad,* The Rotary Club of Ahmedabad, 1949, pp. 12–13.

2.   Dwijendra Tripathi and Makrand Mehta, *The Nagarsheths of Ahmedabad, The History of an Urban Institution in Gujarat City,* Indian History Congress vol. I, Aligarh, 1979, pp. 481–8.
Makrand Mehta, *Business Environment Urbanization and Economic Change in India: A Case Study of Ahmedabad* in the 19th Century in J.S. Grewal and Indu Banga, ed. *Studies in Urban History,* Amritsar, 198, p. 6.

3.   Neera Desai, *Social Change in Gujarat, A Study of Nineteenth Century Gujarat,* Bombay, 1978, pp. 282–5; and Hiralal T. Parikh, *Gujarat Vernacular Society no itihas* (History of Gujarat Vernacular Society) pt. 1, pp. 14–15, 1921, Ahmedabad.

4.   Gillion, *Ahmedabad,* pp. 72 and 77.

5.   Makrand Mehta, *The Ahmedabad Cotton Textile Industry: Genesis and Growth,* Ahmedabad, 1982, p. 7.

6.   Mehta, Business Environment, pp. 5–6.

7.   Gillion, *Ahmedabd,* p. 51.

8.   Gillion, *Ahmedabad,* p. 85; and M.V. Kamath & V.B. Kher, *The Story of Militant But Non-Violent Trade Unionism. A Biographical and Historical Study.* Ahmedabad, 1993, p. 24.

9.   *Representative Men of the Bombay Presidency,* Bombay, 1900, Vol. 3, p. 166.

10.   Gillion, *Ahmedabad,* p. 69.

11.   *Representative Men of the Bombay Presidency,* Bombay, 1900, Vol. 4, p. 205.

12.   *Gautam Sarabhai, Curumchand Premchand,* 1975, p. 2 (for private circulation only).

13.   *Representative men of the Bombay Presidency,* Bombay, 1900, Vol. 4. p. 205.

14.   Kamath and Kher *Non Violent Trade Unionism.* pp. 34 and 35.

15.   Leena Mangaldas. *Akhand Divo,* (The Eternal Lamp), Ahmedabad, 1979, pp. 8–21.

16.   Kamath and Kher, *Non-Vioent Trade Unionism.*

17.   Mangaldas, *Akhand Divo,* Ahmedabad, pp. 16–19.

18. Sarabhai, *Curumchand Premchand,* p. 4
19. E. Williams, 'Forty Years Ago', (unpublished manuscript) p. 24.
20. Mangaldas, *Akhand Divo* p. 21.
21. E. Williams, 'Forty Years Ago' pp. 24 and 26.
22. Williams, 'Forty Years Ago'. p. 25.
23. Mangaldas *Akhand Divo*: p. 51
24. Mangaldas, *Akhand Divo*, p. 18.
25. Shivshankar Shukla, *Anusuyaben Sarabhai ni Jivan Katha*, Ahmedabad, nd., pp. 38–9.
26. M.K. Gandhi, *An Autobiography or the Story of my Experiments with Truth.* Ahmedabad, 1927, p. 297.
27. Ibid. p. 297.
28. Gandhi, *Autobiography,* p. 300.
29. Shankarlal Banker, *Ghandhiji aur Majur Pravrutti* (Gandhiji and the Labour Movement). Ahmedabad, 1965, pp. 86–7.
30. Erik Erikson, *Gandhi's Truth*, London, 1965, p. 307.
31. Kamath and Kher. *Non-Violent Trade Unionism*, pp. 104•34.
32. Banker, *Majur Pravrutti* p. 25.
33. Gandhi, *Autobiography*, pp. 345–6.
34. Gandhi, *Autobiography* pp. 345–6.
35. Mangaldas, *Akhand Divo*, pp. 79–80.
36. William, 'Indian Twilight' (unpublished Manuscript in Sarabhai Foundation).
37. Interview given by Gira Sarabhai to the author.
38. Mangaldas, *Akhand Divo* pp. 32–3.
39. Mangaldas, *Akhand Divo* p. 45.
40. *CWMG*, vol. XXX, p. 267, vol. LXXXVII, pp. 103, 326. vol. LXXXVIII, p. 304.
41. *CWMG*, vol. LXXXVII, p. 23.
42. Erikson, p. 41.
43. Jawaharlal Nehru, *Discovery of India,* Bombay, Asia Publishing House, 1972, p. 358.

II. Satyagraha

1. Standing, 'Indian Twlight', vol. 3, p. 69.
2. Williams, 'Forty Years Ago', p. 11.
3. Standing, 'Indian Twlight', vol III pp. 99–100.
4. Williams, 'Forty Years Ago', p. 61.
5. Ibid. p. 64.
6. M.S. to Ambalal, 16 May 1924.
7. Standing, 'Indian Twilight, pp. 7–8 and 17.
8. Interview given by Puratan Buch to author in Bombay on 13 January 1988.

9.　Puratan Buch, 'Mridula Sarabhai', *Mumbai Samachar,* 2 November 1974.
10.　D.G. Tendulkar, *Mahatma,* vol. III, 1930–34, Bombay, 1953, p. 24.
11.　Erikson, *Gandhi's Truth.* pp. 445–6.
12.　*Jyoti Vikas Yatra,* Jyoti Sangh, Ahmedabad, p. 290.
13.　N.R. Pathak, ed., *Source Material for a History of the Freedom Movement in India.* vol. III, pt. III, p. 31.
14.　*Bombay Chronicle,* 16 April 1930.
15.　*Young India,* 10 April 1930.
16.　*Prajabandhu,* 25 April 1930.
17.　*Times of India,* 10 April 1930, 2 April 1930.
18.　Letter from Jiwanlal Diwan, Secretary, *Gujarat Prantik Samiti* to J.N., 4 June 1930; and File No. G-11p/1930. AICC Papers, N.M.M.L.
19.　Home Political File. 1932 14 April 1932 N.A.I.
20.　Vinodini Nilkanth. *Gujarat Samachar,* 31 August 1984.
21.　*Tribune,* 9 April 1930, 13 April 1930.
22.　Ibid. 15 April 1930.
23.　Mahadev Desai, *Diaries,* vol. 13, Ahmedabad, 1954, p. 303.
24.　Kantilal Shah, *Viramgam Satyagraha.*
25.　Gandhi's statement 7 April 1934 in P. Sitaramayya, *The History of the Indian National Congress,* vol. 1, Delhi, 1935, p. 259.
26.　M.S. to M.G. 9 April 1934.
27.　Mahadev Desai to M.S., 4 Jan 1941.
28.　Purushottam Mehra, *A Dictionary of Modern Indian History 1707–1947,* Delhi, 1985, p. 573.
29.　Nirad C. Chaudhuri, *Thy Hand Great Anarch,* London, Addison Wesley, 1989, pp. 11–2.
30.　M.S. talk on All India Radio, 15 June 1969.

III. In Prison

1.　M.S. Jail Diary, 26 Nov 1930.
2.　M.S. Jail Diary, 31 Jan 1932.
3.　Interview with Devyani Desai, Bombay, 13 Nov 1988. (Interview was in Gujarati. Translation by author).
4.　M.S. Jail Diary.
5.　M.S. Jail Diary, 13 Aug 1933.
6.　Ibid. 25 May 1932.
7.　Ibid. 8 June 1932.
8.　Sardar Patel to Saraladevi, Sardadevi Sarabhai Papers, 25 April 1933.
9.　Jail Diary, 21 June 1932.
10.　Interview given by Kaliprasadji to author.
12.　M.S. Jail Diary, 21 April 1933.
13.　Ibid. 15 July 1933.

14. M.S. Jail Diary, 4 Aug 1933.
15. Ibid. 7 Sept 1933.
16. Ambalal Sarabhai to M.S., 22 March 1934, M.S. Papers.
17. Ambalal Sarabhai to M.S., 4 April 1934, M.S. Papers.
18. M.S. Jail Diary, 23 Aug 1933.
19. Ibid. 2 September 1933.
20. Ibid. 25 May 1932.
21. Ibid. 25 May 1932.
22. Ibid. 29 June 1933.
23. Morarji Desai to M.S., 9 February 1939, M.S. Papers.
24. Subhas Bose to M.S., 16 December 1938, M.S. Papers.
25. J.N. to M.S., 25 December 1939, M.S. Papers.
26. Ambalal Sarabhai to the Superintendent, Rajkot jail, 15 Dec, 1938.
27. H.V.R. Iyengar, *Sunday Standard,* 17 September 1964.
28. Ambalal Sarabhai to M.S., 18 October 1941.
29. Mrinalini Desai, *Gujarat Samachar,* 17 November 1974.
30. M.S. to Ambalal Sarabhai, 15 October 1943.

IV. Indian National Congress

1. Devavrat Pathak and Praveen Seth, *Sardar Vallabhbhai Patel, From Civic to National Leadership,* Navjivan Publishing House, Ahmedabad, 1980, Narhari D. Parikh, *Sardar Vallabhbhai Patel,* vol. I, Navjivan Publishing House, Ahmedabad, 1956; Rajmohan Gandhi, *Patel, A Life,* Navjivan Press, Ahmedabad, 1992.
2. Mahadev Desai, *Vir Vallabhai,* Sardar Patel University, 1974. p. 27.
3. M.S. papers, 9 September 1992.
4. Vanmala Desai, daughter of Narhari Parikh, wife of Mahendra V. Desai, was brought up in the Sabarmati Ashram. She took part in the Quit India Movement. Interview given by her to the author in July 1988, New Delhi.
5. Interview given by Harivadan Thakore to author in Ahmedabad.
6. M.S. Papers 9 September 1972.
7. M.S. to M.K. Gandhi, 2 August 1936.
8. M.S. to Vallabhbhai Patel, n.d.
9. M.S. to Vallabhbhai Patel, 10 January 1946.
10. Ambalal Sarabhi to M.S., 16 April 1992, Ambalal Sarabhai papers.
11. Ambalal Sarabhai to Bharati Sarabhai, 13 February 1939, *ibid.*
12. Vallabhbhai Patel to M.S., nd.
13. Vallabhbhai Patel to M.S, 2 November 1935.
14. Ambalal Sarabhai to M.S., 30 September 1947.
15. M.S. to Vallabhbhai Patel, 27 October 1934.
16. Vallabhbhai Patel to M.S., 18 April 1936.
17. M.S. to Vallabhbhai, 29 April 1936.

18. Morarji Desai, *The Story of My Life*, vol. 1, p. 94.
19. Morarji Desai to M.S., 15 April 1941.
20. Ibid.
21. M.S. to J.N., 27 November 1936.
22. J.N. to M.S., 4 February 1940.
23. M.S. to Vallabhbhai Patel and M.S. to Morarji Desai, 9 February 1946.
24. Vallabhbhai Patel to M.S., 11 February 1946.
25. M.S. to Jairamdas Daulatram, 17 May 1946.
26. M.S. to Sadiq Ali, 11 July 1946. AICC Papers, File G. 18, NMML.
27. M.S. to her family, August, 1946.
28. M.K. Gandhi, *Harijan,* 14 August 1946.
29. Eleanor Morton, *Women Behind Mahatma Gandhi,* Max Reinhardt, London, 1954, p. 199.
30. M.S. to J.B. Kripalani, 24 October 1946.
31. J.B. Kripalani, 24 October 1946.
32. Mahatama Gandhi to Jivanji Desai, 24 August 1944, *CWMG,* vol. LXXVII, pp. 46–7.
33. M.G. to M.S., 25 February 1934.
34. M.G. to M.S., 18 November 1935.
35. M.G. to Miraben, 18 September 1944, *CWMG,* vol. LXXVIII, p. 112.
36. Entry dated 8 August 50 in Diary of Vallabbhai Patel quoted in Rajmohan Gandhi, *Patel,* p. 522.
37. M.S. to Kantilal Ghia, M.S. to Morarji Desai, M.S. to Pattabhai Sitaramayya, 9 September, 1950, 12 Sept., 1950.
38. *Times of India,* 23 May, 1958.
39. Resolution passed by GPCC, 28 June 1958.
40. Ibid.
41. M.S. to AICC, 17 July 1958.
42. Ibid.
43. M.S. to AICC, 14 October 1959.

V. Towards Women's Equality

1. M.S. Jail Diary, 30 March 1932.
2. Ibid.
3. Marie Seton, *Panditji,* London, 1967 p. 114. Another version of the story is that Nehru said: 'Come in the dress you are in that itself is fancy dress.'
4. M.G. to M.S. quoted in Susmita Medh, *Vaitalik,* Ahmedabad, 1987, p. 69.
5. M.G. speech at foundation-stone laying ceremony of *Jyoti Sangh.* 29 June 1934, *Jyoti Sangh* Report, 1936–7, p. 1.
6. *Jyoti Sangh* Report, 1936–7, p. 3.
7. Susmita Medh, *Vaitalik,* Ahmedabd, 1987, p. 103.

8. Interview of Vishakhaben Mehta, 11 September 1976; 14 September 1976.
9. Susmita Medh, *Vaitalik*, Ahmedabad, 1987, p. 97.
10. M.G. to M.S., 7 August 1946.
11. M.S. to Motilal Veen, 27 January 1942.
12. M.S. to J.N., 25 April 1936.
13. Ibid.
14. Ibid.
15. From Gujarat Congresswomen to J.N., nd 1936. Also *CWMG*, Appendix IV, vol. LXII, letter from J.N. to M.G., 25 May 1936.
16. J.N. to M.S., 5 May 1936.
17. Ibid.
18. J.N. to M.G., 5 July 1936, *SWJN* vol. 7, p. 312.
19. M.G. to J.N., 29 May 1936 *CWMG* vol. LXII, Item no. 534.
20. *SWJN*, vol. 7, p. 313. Statement printed in *Bombay Chronicle,* 7 July 1936.
21. J.N. to M.S., 23 February 1940.
22. Ranti Rajwade to M.S., 19 September 1940.
23. M.S. to J.B. Kripalani, 5 July 1939.
24. M.S. to J.B. Kripalani, 10 August 1939.
25. M.S. to Vijayalakshmi Pandit, 31 December 1939.
26. AICC Papers, File No. G. 50.
27. Ibid.
28. Rajendra Prasad to M.S., 19 February 1940.
29. M.S. to Maulana Azad, 14 April 1940.
30. Sucheta Kripalani to M.S. 1940.
31. M.S. to J.B. Kripalani, 19 January 1946, AICC file No. G. 22, NMML.
32. M.S. to J.B. Kripalani, 13 March 1936.
33. M.S. to Umabai Kundapure, Dr. Hardekar, etc., 3 June 1939.
34. M.S. General letter, 5 July 1939.
35. M.S. to Durgabai Jog, 8 August 1939.
36. Aparna Basu and Bharati Ray, *Women's Struggle: A History of All India Women's Conference, 1927–87*, New Delhi, 1990.
37. M.G.'s speech, 11 April 1945. *CWMG*, vol. LXXIX, p. 365.
38. Interview of Author with Pupul Jayakar, August 1988, New Delhi.
39. M.G. to A.V. Thakkar, 1 June 1945, ibid., p. 312.
41. M.G. to M.S., 29 April 1945, M.S. Papers.
42. M.G. to M.S., 15 October 1945, *CWMG,* vol. LXXXI, pp. 357–8.
43. M.G. to Anasuya Sarabhai, 2 September 1945, ibid., p. 211.
44. M.S. to M. Chalapathi Rau, 22 January 1949.

VI. Communal Riots

1. The *Khaksar* (humble) Movement (1938–63) was organised by *Inayat-Ullah Kahn Mashrijqi* in 1931 among lower middle class Muslims. Its members underwent military drill in uniform in order to become soldiers of Islam.
2. M.S. to M.G. 29 April 1941, File no. 10/B/2 p. 1.
3. M.S. to Vallabhbhai Patel, 17 April 1941.
4. M.S. to M.G., 29 April 1941, File no. 10/B/2, pt I.
5. M.S. To J.N. 16 June 1941, File no. 10/B/2 pt I
6. Statement of Major General Shah Nawaz Khan, AICC, File No. 20/1846.
7. Shah Nawaz Khan to M.A. Jinnah. 13 Nov. 1946, AICC File no. 20/1846.
8. Notes issued by M.S. 13 November 46. AICC papers. 20/1846.
9. *CWMG* vol. LXXXVI, p. 283.
10. Pyarelal, *Mahatma Gandhi, the Last Phase,* vol. I, book 2, p. 269.
11. Ibid.
12. Ibid. p. 288.
13. M.G. to Amrit Kaur, *CWMG* vol. LXXXVII, p. 496.
14. M.S. *Pratham Pratyaghat* Ahmedabad, 1989, pp. 27–8; Songs to Tagore rendered into English by Aruna Chakravarti, Bombay. *Vaitalik.* nd.
15. M.G.'s talk with Muslim refugees at Hansdih, 21 March 1947, *CWMG.* vol. LXXXVII, p. 132.
16. Ghanshyam Shah, 'Communal Riots in Gujarat: Report of a preliminary investigation', *Economic and Political Weekly,* vol. nos. 3, 4, 5, Annual Number January, 1970.
17. Ibid.
18. According to another estimate, 512 persons were killed of whom 413 were Muslims. A.R. Desai, 'Caste and Communal Violence in Post-Partition Indian Union', in Asghar Ali Engineer ed., *Communal Riots in Post-Independence India,* New Delhi, 1984, p. 22.
19. M.S. to Balbheri and M.S. to Jayprakash Narayan, 22 October 1969.
20. A.M. Peerzada to M.S. 27 March 1970.
21. Shah Nawaz Khan to M.S. 22 May 1971.
22. Khalil Ahmed to M.S. 2 June 1971.
23. Abdul Ghaffar Khan's speech to Paliament, 24 November 1969.
24. M.S.'s note on Khan's reaction, 11 June 1970.
25. Ibid.
26. M.S. to Jayaprakash Narayan, 15 September 1971.
27. M.S. to Shah Nawaz Khan, 21 September 1971.
28. M.S. to Kunwar Mohinder Singh Bedi, 25 September 1971.

VII. The Great Migration

1. Kripal Singh, ed., Select Documents on Partition of Punjab, New Delhi, p. XXIX.
2. Horace Alexander, *New Citizens of India*, OUP, p. 7, (quoted in Kripal Singh).
3. Francis Moodie to M.A. Jinnah, 5 September 1947 (quoted in Kripal Singh).
4. M.S. to Ambalal Sarabhai and family, 18 August 1947.
5. K. Iftikar, *Jab Amritsar Jal Raha Tha* (When Amristsar was Burning), 9th edition, Lahore, 1991, pp. 259–61.
6. Kartar Singh Duggal, *Twice Born, Twice Dead*, New Delhi, 1979, p. 130.
7. H.V.R. Iyenger, *Sunday Standard,* 27 September 1964.
8. Lady Edwina Mountbatten to M.S. 8 November 1947.
9. *SWJN*, vol. 5, second series, p. 195.
10. M.S. to Gopalaswamy Iyengar, 30 November 1947.
11. J.N. to Mohanlal Saxena, 8 June, 1948, *SWJN*, vol. VI, p. 124.
12. *The Hindu.* 16 Nov. 1947.
13. Chandulal Trivedi to Swaran Singh, 11 Aug. 1947, 12 August 1947. File D/269 Broadland Archives.
14. Francies Moodie to Chandulal Trivedi, 17 August 1947, MSS Evr. F. 641/17.
15. Swaran Singh to Chandulal Trivedi. 17 Aug. 1947, MSS Evr. F. 641/17.
16. J.N. to Gopichand Bhargava, 14 June, 1948, *SWJN*, vol. VI, pp. 126–7.
17. Interview with L.C. Jain. who was involved in the rehabilitation of refugees in Faridabad with Kamaladevi Chattopadhyaya.
18. The Hidden Hand-II, *Bhoodan.* 24 June 1961.
19. M.S. to Rajendra Prasad, 5 February 1955.
20. Rajendra Prasad to M.S., 18 May, 1955.

VIII. Recovery of Abducted Women

1. Anees Kidwai, *Azadi Ki Chhaon Mein*, luen New Delhi, 1990, p.
2. 'Evacuation Report of Dera Gazi District', LAR File no. LIX 19.
3. Report of work in 'Gujaranwala District', LAR File no. LIX 12.
4. Kidwai, *Azadi ki Chhaon Mein.*
5. Y.D. Gundevia, *Outside the Archives*, Delhi, 1984, pp. 35, 38.
6. M.S. Note on Purchase of Abducted Women in Sargoda district (W. Punjab), 27 Nov. 1947, M.S. Papers.
7. Lady Edwina Mountbatten's note of 28 Nov. 1947, ibid.
8. *CWMG* vol. XC p. 128.
9. Steering Committee meeting for Recovery of Abducted Women, MEO. Records no. F 10899/H vol. II.

10. *CWMG* vol. XC, p. 194.
11. M.S. The Recovery of Abducted Women article given to national Press Syndicate to be released on 25 July 1948, M.S. Papers.
12. M.S. to Gopalaswamy Iyengar, 8 July 1948.
13. Ibid. and Gundevia *Outside the Archives* p. 39
14. M.S. to Gopalaswamy Iyengar, 8 July 1948.
15. Gundevia, *Outside the Archives* pp. 40–1.
16. Ibid. pp. 40–1.
17. Ibid. p. 41.
18. Ibid. p. 43.
19. Ibid. pp. 44–5
20. Kamlaben Patel, *Mul Suta Ukhdela* (The Uprooted),
21. Rameshwari Nehru Papers, Memorandum on Recovery of Women 20 June 1949. NMML 19.
22. Ibid.
23. Gundevia to M.S.

IX. Kashmir

1. Karan Singh, *Heir Apparent* New Delhi, 1982, page 156.
2. Karan Singh, *Heir Apparent*. New Delhi, 1982, pp. 158–9.
3. Ibid. 156.
4. B.M. Kaul, *The Untold Story* Bombay, 1967, pp. 137–47.
5. Ibid.
6. Ibid. p. 146.
7. Frank Moraes, *Witness to An Era*. Bombay, 1973, pp. 219–20.
8. B.N. Mullik. *Kashmir*. Bombay, 1971, p. 67.
9. M.S. to Sheikh Abdullah, henceforth referred to as S.A. 1 Dec. 1953.
10. S.A. to M.S. 9 December 1953.
11. S.A. to M.S. 10 March 1954.
12. Farooq Abdullah to M.S. from Medical College Hospital, Jaipur, nd.
13. Farooq Abdullah to M.S. 4 October 1964.
14. Farooq Abdullah to M.S. 27 August 1960.
15. Begum Abdullah to M.S. 2 June 1961.
16. Interview of Begum Abdullah with the author, 28 June 1988.
17. Ibid.
18. Farooq Abdullah to the author in an interview, 29 June 1988.
19. G.M. Shah in a letter to the author, 19 July 1966.
20. Letter from Sadr-Ud-Din Mujaheed to the author, 6 July 1988.
21. Telegram from M.S. to Mubarak Shah, 10 November 1966.
22. S.A. to M.S., 6 June 1955.
23. M.S. to Rajendra Prasad, 11. October 1955.
24. S.A. to M.S., 2 July 1956.
25. S.A. to Sadiq, 26 September 1956.

26. Sadiq to S.A.
27. S.A. to Sadiq, constitution of Kashmir.
28. *Tribune.* 8 August 1958
29. M.S. Notes
30. *Lok Sabha Debates.* J.B. Kripalani's speech, 8 December 1958.
31. Ibid., J.N.'s speech, 9 December 1958.
32. M.S. Note written in Tihar Jail.
33. A.C. Sen, interview with author, 8 February 1988.
34. Amnesty International Bulletin No. 15 May 1966.
35. Ibid.
36. M.S. to Rao Saheb Patwardhan, 29 June 1967.
37. M.S. to Lal Bahadur Shastri, 11 September, 1965
38. M.S. to Y.B. Chavan, 26/28 November, 1966.
39. M.S. to G.L. Nanda, 3/5 October, 1966 and 3 November 1966
40. J.B. Kripalani's speech in the Lok Sabha, 21 November 66.
41. Andrew Mann to M.S. 31 December 1965.
42. Elizabeth Gordon to M.S. 24 March 1966.
43. Telegrams, S.M. Joshi to M.S., Humayun Kabir to M.S., P.N. Sapru to M.S.

X. The Crusader

1. A.R. Desai, *Social Background of Indian Nationalism,* Popular Prakashan, Bombay, 1975, pp. 211–2.
2. *SWMG,* vol. VI, pp. 482–9.
3. Vina Mazumdar, 'The Social Reform Movement in India: From Ranade to Nehru' in B.R. Nanda, ed, *The Indian Women: From Purdah to Modernity,* Vikas, Delhi, 1976; and Aparna Basu, 'Role of Women in the Indian Freedom Struggle' in Nanda, *Indian Women.*
4. Madhu Kishwar, 'Gandhi on Women', in *Economic and Political Weekly,* vol. 20, no. 40, 5 October 1985, p. 1698.
5. Lakshmi Menon, in Devaki Jain, ed., *Indian Women,* New Delhi, 1975, p. 23; and Kamala Devi Chattopadhyaya, *Indian Women's Battle for Freedom,* New Delhi, 1983.
6. Madhu Kishwar, 'Gandhi on Women'.
7. Mridula Sarabhai to Vinoba Bhave, 9 March 1969 (personal).
8. Sarladevi to Indira Gandhi, 24 December 1974.
9. Interview of A.C. Sen with author.
10. Interview of Farooq Abdullah with author.
11. Indira Gandhi to Sarladevi Sarabhai, 29 October, 1974.
12. Apa Pant to Bharati Sarabhai, 2 November, 1974.

BIBLIOGRAPHY

Files and Manuscripts

All India Congress Committee files and manuscripts (N.M.M. & L)
Home Political Files (N.A.I.)
Ambalal Sarabhai Papers (Sarabhai Foundation)
Mridula Sarabhai Papers (Sarabhai Foundation)
Sarladevi Sarabhai Papers (Sarabhai Foundation)
Rameshwari Nehru Papers (N.M.M. & L)
Gautam Sarabhai, Currumchand Premchand (Sarabhai Foundation)
Standing, William, 'Indian Twilight' (Sarabhai Foundation)
Williams, E. 'Forty Years Ago' (Sarabhai Foundation)

Newspapers

Bombay Chronicle
Gujarat Samachar
Hindu
Janmabhoomi
Mumbai Samachar
Prajabandhu
Sandesh
Sunday Standard
Times of India
Tribune

Journals

Amnesty International Bulletin
Bhoodan
Economic and Political Weekly
Illustrated Weekly of India
Jyoti Vikas Yatra
Young India

Books/Articles

Ahmedabad. The Rotary Club of Ahmedabad, 1949.

Banker, Shankerlal. *Gandhiji aur Majur Pravritti,* (Gandhiji and The Labour Movement), Ahmedabad, 1965.

Basu, Aparna and Ray, Bharati. *Women's Struggle: A History of All India Women's Conference, 1927–87,* Manohar, New Delhi, 1991.

Bhatt, Chhotubhai. 'Mridula Sarabhai: Jivan Jhanki' (Mridula Sarabhai: A Life Sketch), *Kasturba Darshan,* Ahmedabad, 1975.

Desai, A.R. 'Caste and Communal Violence in Post Partition Indian Union' in Ashghar Ali Engineer, ed., *Communal Riots in Post Independence India,* New Delhi, 1984.

Desai, Mahadev, *The Diary of Mahadev Desai, Vol. I,* Navjivan Press, Ahmedabad, 1953.

Desai, Mahadev. *Vir Vallabhbhai,* Sardar Patel University, Vallabh Vidyanagar, 1974.

Desai, Morarji. *The Story of My Life,* Macmillan, New Delhi, 1974.

Desai, Neera. *Social Change in Gujarat, A Study of Nineteenth Century Gujarat Society.* Bombay, Vora & Co., 1978.

Erikson, Erik. *Gandhi's Truth,* W.W. Norton, New York, 1969

Gandhi, Mohandas Karamchand. *Autobiography, or the Story of My Experiments With Truth,* Ahmedabad, 1927.

Gandhi, Mohandas Karamchand. *Collected Works of Mahatma Gandhi,* Navjivan, Ahmedabad, 1992.

Gandhi, Rajmohan. *Patel, A Life,* Navjivan, Ahmedabad, 1992.

Gillion, Kenneth. *Ahmedabad: A Study in Indian Urban History,* Berkeley, 1968.

Gordon, A.D.D. *Businessmen and Politics: Rising Nationalism and a Modernizing Economy in Bombay, 1918–1933,* New Delhi, 1978.

Gundevia, Y.D. *Outside the Archives,* Delhi, 1984.

Kamath, M.V. and Kher, V.B. *The Story of Militant But Non-Violent Trade Unionism, A Biographical and Historical Study,* Ahmedabad, 1993.

Singh, Karan. *Heir Apparent,* New Delhi, 1982.

Kaul, B.M. *The Untold Story,* Bombay, 1967.

Kidwai, Anees. *Azadi Ki Chhaon Mein* (In the Shadows of Independence) Tr. by Noor Nabi Abbasi, New Delhi, 1990.

Singh, Kirpal, *Partition of the Punjab,* Patiala, 1972.

Mangaldas, Leena. *Akhand Divo* (The Eternal Lamp), Ahmedabad, 1979.

Medh, Susmita. *Vaitalik,* Ahmedabad, 1987.

Mehra, Purushottam. *A Dictionary of Modern Indian History, 1707–1947,* Delhi, 1985.

Mehta, Usha. 'Mridula Sarabhai', News Service, AIR, New Delhi.

Mehta, Makrand. 'Business Environment, Urbanization and Economic Change in India: A Case Study of Ahmedabad in the 19th Century' in J.S. Grewal and Indu Banga ed., *Studies in Urban History,* Amritsar, 1981

Mehta, Makrand. *The Ahmedabad Cotton Textile Industry: Genesis and Growth*, Ahmedabad, 1982.

Minault Gail. *The Extended Family: Women and Political Participation in India and Pakistan*, Delhi, 1984.

Moraes, Frank. *Witness An Era*, Bombay, 1975.

Morton, Eleanor. *Women Behind Mahatma Gandhi*, London, 1954.

Mullick, B.N. *Kashmir*, Bombay, 1974.

Nehru, Jawaharlal. *Selected Works of Jawaharlal Nehru* (15 vols), ed. S. Gopal, Orient Longmans, Delhi.

Pandey, B.N. *Nehru*, Macmillan, London, 1976.

Parikh, Hiralal. *Gujarat Vernacular Society no Itihas* (History of Gujarat Vernacular Society), Ahmedabad.

Parikh, Narhari D. *Sardar Vallabhbhai Patel*, Ahmedabad, 1956.

Pathak, Devavrat and Seth, Praveen. *Sardar Vallabhbhai Patel, From Civic to National Leadership*, Ahmedabad, 1980.

Pathak, N.R. ed. *Source Material for a History of the Freedom Movement in India*, vol. III, pt. III.

Patel, Kamalaben. *Mul Sotan Ukhdelan* (uprooted), Ahmedabad, 1985.

Pyarelal. *Mahatma Gandhi, The Last Phase*, Navjivan, Ahmedabad, vol. 1, book 2.

Representative Men of the Bombay Presidency, vols 3 & 4, Bombay, 1900.

Sarabhai, Mridula. *Pratham Pratyaghat* (First Assault), Ahmedabad.

Seton, Mary. *Panditji*, London, 1967.

Shah, Ghanashyam. 'Communal Riots in Gujarat: Report of a Preliminary Investigation', *Economic and Political Weekly*, vols 3, 4, 5. Annual Number, Jan. 1970.

Sheikh, Abdullah. *Aatish-e-Chinar* (Flames of the Chinar), Srinagar, 1986.

Sitaramayyah, Pattabhi. *The History of the Indian National Congress*, vol. 1, Delhi, 1935, vol. II, Bombay, 1947.

Shukla, Shivshankar. *Ansuyaben Sarabhai ni jivan katha* (Biography of Anasuya Sarabhai), Ahmedabad, n.d.

Tendulkar, B.G. *Mahatma, Life of Mohandas Karamchand Gandhi, vol. III*, Bombay, 1953.

Tripathi, Dwijendra and Mehta, Makrand. 'The Nagarsheths of Ahmedabad, The History of an Urban Institution in a Gujarat City, *Indian History Congress Proceeding, vol. 1*, Aligarh, 1979.

Tripathi D. and Mehta M. 'Class Character of the Gujarati Business Community in D. Tripathi, ed. *Business Community in India*, New Delhi, 1984.

INDEX